Principles of

Plant Pathology

Principles of

Plant Pathology

Prof. P. VIDHYASEKARAN
Director
Centre for Plant Protection Studies
Tamil Nadu Agricultural University
Coimbatore - 641 003
(India)

CBSPD

CBS Publishers & Distributors Pvt Ltd

New Delhi • Bengaluru • Chennai • Kochi • Kolkata • Lucknow • Mumbai
Hyderabad • Jharkhand • Nagpur • Patna • Pune • Uttarakhand

Principles of
Plant Pathology

ISBN: 978-81-239-0198-5

First Edition: 1993
Reprint: 2001, 2006, 2010, 2012, 2014, 2016, 2017, 2019, 2023

Published by **Satish Kumar Jain** and produced by **Varun Jain** for
CBS Publishers & Distributors Pvt Ltd
4819/XI Prahlad Street, 24 Ansari Road, Daryaganj, New Delhi 110 002, India
Ph: 011-23289259, 23266861, 23266867 Website: www.cbspd.com
Fax: 011-23243014 e-mail: delhi@cbspd.com

Corporate Office: 204 FIE, Industrial Area, Patparganj, Delhi 110 092, India
Ph: 011-4934 4934 Fax: 011-4934 4935 e-mail: publishing@cbspd.com; publicity@cbspd.com

Branches

- **Bengaluru:** Seema House 2975, 17th Cross, KR Road, Banasankari 2nd Stage, Bengaluru 560 070, Karnataka, India
  Ph: +91-80-26771678/79 Fax: +91-80-26771680 e-mail: bangalore@cbspd.com
- **Chennai:** 7, Subbaraya Street, Shenoy Nagar, Chennai 600 030, Tamil Nadu, India
  Ph: +91-44-26680620, 26681266 Fax: +91-44-42032115 e-mail: chennai@cbspd.com
- **Kochi:** 42/1325, 1326, Power House Road, Opp KSEB, Power House, Ernakulum Kochi 682 018, Kerala, India
  Ph: +91-484-4059061-65,67 Fax: +91-484-4059065 e-mail: kochi@cbspd.com
- **Kolkata:** 147, Hind Ceramics Compound, 1st Floor, Nilgunj Road, Belghoria, Kolkata-700056, West Bengal, India
  Ph: +033-25633055, 033-25633056 e-mail: kolkata@cbspd.com
- **Lucknow:** Basement, Khushnuma Complex, 7 Meerabai Marg (Behind Jawahar Bhawan),Lucknow-226001, UP, India
  Ph: +0522-4000032 e-mail: tiwari.lucknow@cbspd.com
- **Mumbai:** PWD Shed, Gala no 25/26, Ramchandra Bhatt Marg, Next to JJ Hospital Gate no. 2, Opp. Union Bank of India, Noorbaug, Mumbai-400009, Maharashtra, India
  Ph: 022-66661880/89 e-mail: mumbai@cbspd.com

Representatives

- Hyderabad 0-9885175004
- Jharkhand 0-9811541605
- Nagpur 0-9421945513
- Patna 0-9334159340
- Pune 0-9923910676
- Uttarakhand 0-9716462459

Printed at Glorious Printers, Jhilmil Industrial Area, Delhi, India

FOREWORD

In recent times, 'Plant Pathology' as a science has made rapid progress in many facets. The mechanisms of pathogenesis of fungi, bacteria, viruses, viroids, virusoids, mycoplasma-like organisms, spiroplasma, rickettsia-like organisms and algae in crop plants have been studied in depth to develop very effective management practices against these pathogens. Molecular biology of these organisms is another fast developing field. Disease forecasting technology using computer models is being developed against many important diseases. Biology control of the diseases is also gaining much importance particularly the root diseases. Significant contributions have been made all over the world in evolving disease resistant varieties of crops. Genetic engineering, gene cloning, coat protein gene and tissue culture are the key words of late in developing disease-resistant cultivars.

I am very happy to note that Dr. P. Vidhyasekaran has written the book covering all the modern aspects of Plant Pathology. The advancement made in the science has been rendered in a very simple style. I am quite sure that this book will be highly useful to all the students who are introduced into this fascinating subject as well as to the post-graduate students who specialize in the field of Plant Pathology asw their professional career.

I congratulate Dr. P. Vidhyasekaran for bringing out this valuable book.

Tamil Nadu Agrl. University
Coimbatore-641003
April 4, 1992

S. JAYARAJ
Vice-Chancellor

PREFACE

The subject 'Plant Pathology' has made a marvellous progress during the past five years. Previously plant pathologists used to have razor blades, take sections and study the morphology of fungi and life cycle of some of the fungal pathogens, has also been studied. Now, plant pathologists contribute more and more on mechanism of pathogenesis of all kinds of plant pathogens. Physiological and Molecular Plant Pathology is developing as a separate branch of the science. Computer modelling of epidemics has been achieved for many diseases. Novel techniques are being exploited to evolve resistant varieties. Rate-reducing resistance, durable resistance, pyramiding genes and exploiting gene blocks, strong genes and multiline cultivars, are all newer approaches. Breeding for subsistence agriculture has become a newer dimension. Biological control which was a laboratory study in yesteryears has become a reality in farmers' fields. Transgenic plants with built in disease resistance are being made available. The Universities are revising the syllabi of their graduate courses to suit the modern developments. The book has been written in a simple style covering all the aspects of modern plant pathology to cater to the needs of the undergraduate [B.Sc. (Ag.)] students and to the Botany graduates who opt to specialize in Plant Pathology. This book will also be useful to the post-graduate students as a source book. It will serve as a guide for students appearing for National Entrance Test.

I acknowledge with thanks the help rendered by Mr. G. Umapathi, Assistant Professor of our University, in the preparation of line drawings. My special thanks are due to Dr. S. Jayaraj, Vice Chancellor of our University for providing the Foreword.

P. VIDHYASEKARAN

CONTENTS

1

PLANT PATHOLOGY — A FASCINATING FIELD OF SCIENCE

Plant Pathology is the science which deals with plant diseases. Any impairment of normal physiological functions affecting an organism is called disease. Plants are affected by many diseases caused by fungi, bacteria, viruses, mycoplasma - like organisms, spiroplasmas, rickettsia - like organisms, algae, phanerogamic parasites and nematodes. Nutritional deficiency also leads to diseases.

Crop losses due to diseases are very heavy and many famines have been reported in the history due to crop diseases alone. In 1845 Ireland faced a big famine in which several thousands of people died due to starvation and the cause for the famine had been traced to be due to severe outbreak of late blight of potato. India also faced a similar famine in 1943 and in Bengal many starvation deaths have been reported. The Bengal famine has been reported to be due to brown spot of rice.

Crop diseases have shattered the economy of many countries. In 1800s Sri Lanka was the country which produced maximum quantity of coffee in the world and the country prospered well. But in 1867 coffee rust entered into the country and by 1893 the country's export of coffee fell down by more than 90 per cent. Brazil became the biggest coffee exporter and this country is free from the disease. Similarly wine industry in France was shattered in 1875 due to severe outbreak of downy mildew in grapevine. In 1885, Bordeaux mixture was discovered by Prof. Millardet of Bordeaux University and it effectively controlled downy mildew. This fungicide revived wine industry in France.

Several high-yielding cultivars failed miserably in the farmers' fields because of their susceptibility to diseases. The rice cultivar IR50 is considered as wonder rice yielding 15 to 18 tonnes per ha in Tamil Nadu. But this rice could not be cultivated in the major monsoon season due to severe outbreak of blast. Similarly the rice cultivars ADT 27 and ADT 33 disappeared due to their high susceptibility to bacterial blight. The hybrid bajra HB3 had to be replaced due to occurrence of green ear disease. Hill banana, 'Virupakshi' has almost disappeared in Palani hills (Tamil Nadu) due to banana bunchy top virus.

The following are some of the most threatening diseases which inflict heavy losses to the farmers year after year.

Rice	: blast, bacterial blight, tungro and sheath blight
Wheat	: rusts, smut and karnal bunt
Bajra	: ergot and downy mildew
Jowar	: sugary disease and downy mildew
Groundnut	: rust and late leaf spot (tikka)

Gingelly	:	phyllody
Coconut	:	Thanjavur wilt, Kerala wilt
Sugarcane	:	red rot
Banana	:	panama wilt, bunchy top and sigatoka
Grapevine	:	downy mildew
Potato	:	late blight
Tomato	:	tomato spotted wilt
Betelvine	:	wilt
Pepper	:	quick wilt
Cardamom	:	katte, pod rot

Management of crop diseases is a challenging task. Chemicals to combat fungal diseases alone have been identified and they also do not control soil-borne fungal pathogens effectively and soil-borne diseases are the major ones. Antibiotics developed against human pathogens are being used to control plant bacterial pathogens. The results are not encouraging and possible danger of development of antibiotic-resistant strains of bacteria causing diseases in mammals due to indiscriminate use of the antibiotics exists. No useful chemicals with antiviral properties have been identified. Although mycoplasma - like organisms (MLO) respond to tetracyclines, management of plant diseases caused by MLO with tetracyclines is only fictious and impracticable.

Modern plant pathologists approach the problem by different methods. Resistant cultivars have been developed against many diseases. Vertical and horizontal resistance, partial and rate - reducing resistance, durable resistance, tolerance, strong gene, weak gene, and multiline cultivars are all being exploited to manage diseases. Biological control is also attempted by the plant pathologists. Soil fungistasis, suppressive soil, siderophores, bacteriocin, antagonistic organisms, hyperparasites and cross protection are also studied to use them in the control of diseases. Genetic engineering and tissue culture are the latest biotechnological tools used by the plant pathologists to manage crop diseases.

The book describes all the principles of plant pathology so that the students can take up the challenging task confidently and shine well as plant pathologists or plant protection specialists.

2

THE PATHOGENS - FUNGI

Fungi are the major groups of pathogens causing many severe diseases in plants. The fungi belong to the Kingdom Myceteal.

VEGETATIVE STRUCTURES

The vegetative body of the fungi is called mycelium. *Mycelium* is the mass of hyphae constituting thallus. *Hyphae* are the filamentous structures. A hypha is made of a thin, transparent, tubular wall filled or lined with a layer of protoplasm.

SEPTA

The cross-walls which divide the hypha into cells are called septa. Hyphae of some fungi do not have septa and they are called aseptate or *coenocytic.*

PLECTENCHYMA

During certain stages of fungal development, the mycelium becomes organized into loosely or compact woven tissues, as against the loose hyphae ordinarily found in the mycelium. The organized fungal tissues are called plectenchyma.

PROSENCHYMA

The loosely woven tissue in which the component hyphae with elongated cells lie more or less paralell to one another is called prosenchyma.

PSEUDOPARENCHYMA

The fungal tissues which are closely packed, in the form of more or less isodiametric or oval cells resembling the parenchyma cells of higher plants are called pseudoparenchyma. Both prosenchyma and pseudoparenchyma compose various types of vegetative (somatic) and reproductive structures. Stroma is usually made up of prosenchyma while sclerotium is made up of pseudoparenchymatous tissue. Both stromata (plural of stroma) and sclerotia are somatic structures of fungi.

STROMA

It is a compact, somatic structure which looks like a mattress. On or in the stroma fructifications (the structures containing spores) are formed.

SCLEROTIUM

It is a hard and compact vegetative resting structure resistant to unfavourable conditions. It is mostly made up of pseudoparenchymatous tissue.

RHIZOMORPH

The mycelium of some fungi forms thick strands. In such strands the hyphae lose their individuality and form complex tissues. The strands are called rhizomorph (root like structures). The strand has a thick hard cortex and a growing tip. It looks like a root tip (Fig. 1).

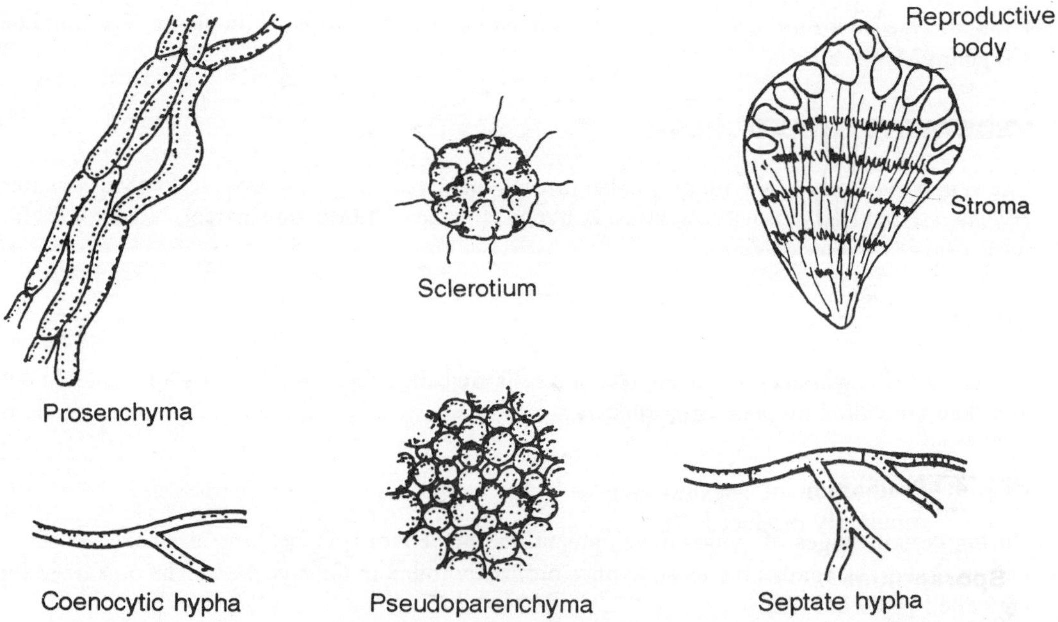

Fig. 1 Vegetative structures of fungi

Asexual Reproduction: Fungi reproduce both asexually and sexually. Asexual reproduction is repeated several times in the life cycle of a fungus while sexual reproduction occurs only once in its life cycle. There are different types of asexual reproduction found in fungi.

1. **Fragmentation of hyphae:** The hyphae break up into their component cells which behave like spores. They are called *oidia* or *arthrospores.* When the cells become enveloped in a thick wall before they separate from each other or from other hyphal cells adjoining them, they are called *chlamydospores.* (Fig. 2)

2. **Fission:** The cell is split into two daughter cells by constriction and formation of a cell wall. This method is called fission which is commonly found in yeasts.
3. **Budding:** Small outgrowth (bud) is produced from a parent cell. The nucleus of the parent cell divides and one daughter nucleus migrates into the bud. The bud increases in size, utimately breaks off and forms a new individual.

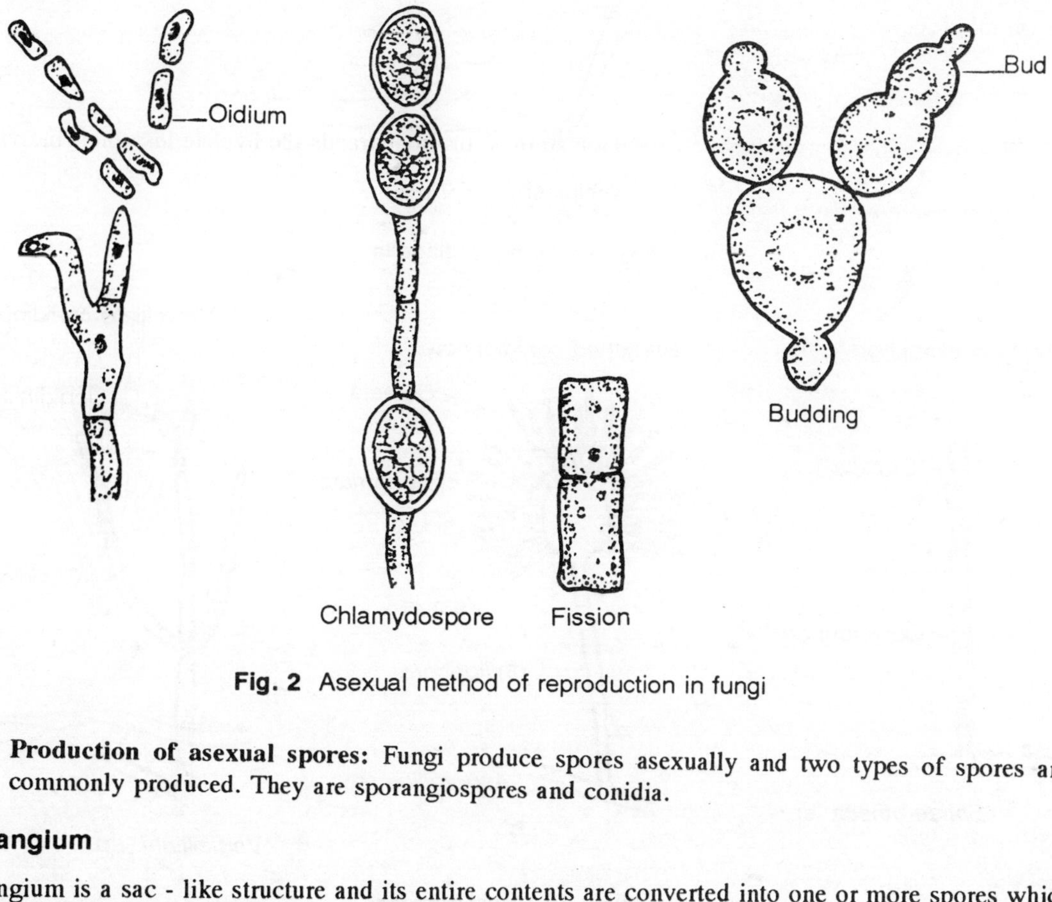

Fig. 2 Asexual method of reproduction in fungi

4. **Production of asexual spores:** Fungi produce spores asexually and two types of spores are commonly produced. They are sporangiospores and conidia.

Sporangium

Sporangium is a sac - like structure and its entire contents are converted into one or more spores which are called *sporangiospores.* If the sporangiospores are motile, they are called *zoospores.* The zoospores may have one or two *flagella.* Two types of flagella are found. One is *Whiplash.* The basal portion of the whiplash is rigid while the terminal portion is flexible. The *tinsel* flagellum is like a feather (Fig. 3).

The non-motile spores are called *aplanospores.*

Conidia

Conidia are produced at the tips or sides of hyphae. The specialized hypha which bears conidia is called *conidiophore.* Conidia are generally borne on conidiophores which may be produced loosely and indiscriminately by the somatic hyphae or grouped in various types of asexual fruiting bodies.

Condiophores may be simple or branched. They may look like somatic hyphae or they may be provided with sterigmata (small hyphal branches which support the conidia) or specialized branches on which they bear the conidia (Fig. 4).

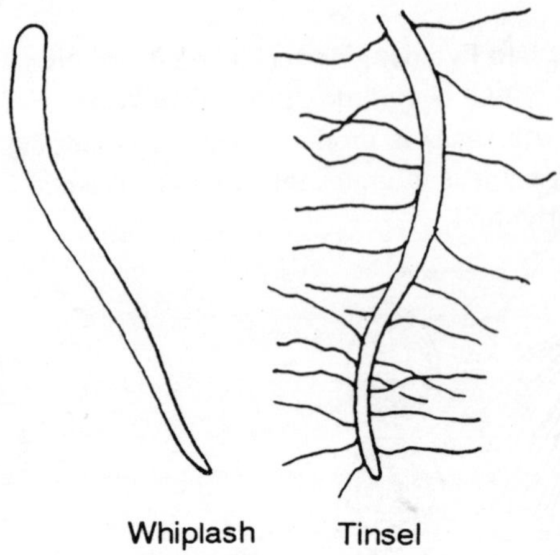

Fig. 3 Types of flagella

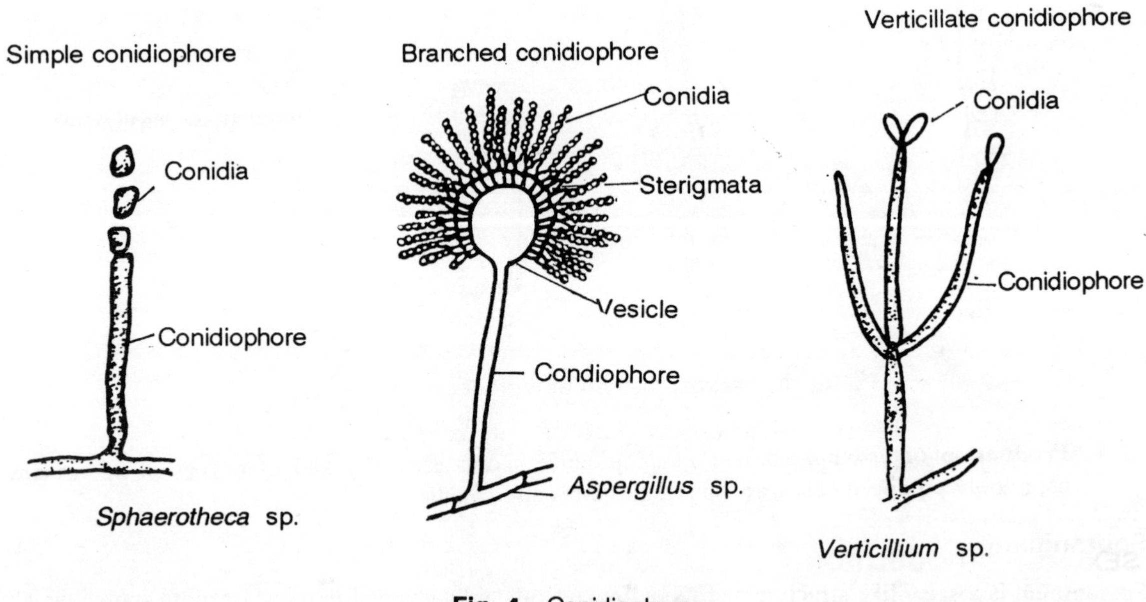

Fig. 4 Conidiophores

Conidiophores may be organized into definite fruiting bodies in case of some fungi.

The various fruiting bodies are:

Pycnidium: It is a hollow, globose or flask shaped structure whose pseudoparenchymatous walls are lined with conidiophores.

Acervulus: Below the epidermis of the host a mat of hyphae is seen. From the mat short conidiophores which are closely packed arise forming a bed-like mass.

Sporodochium: The conidiophores are cemented together to form sporodochium. In this, conidiophores arise from the surface of a cushion-shaped stroma.

Synnema: Synnema (plural synnemata) is a group of conidiophores cemented together to form an elongated, spore-bearing structure. This may be split in different ways near the apex, sometimes resembling a feather duster (Fig. 5).

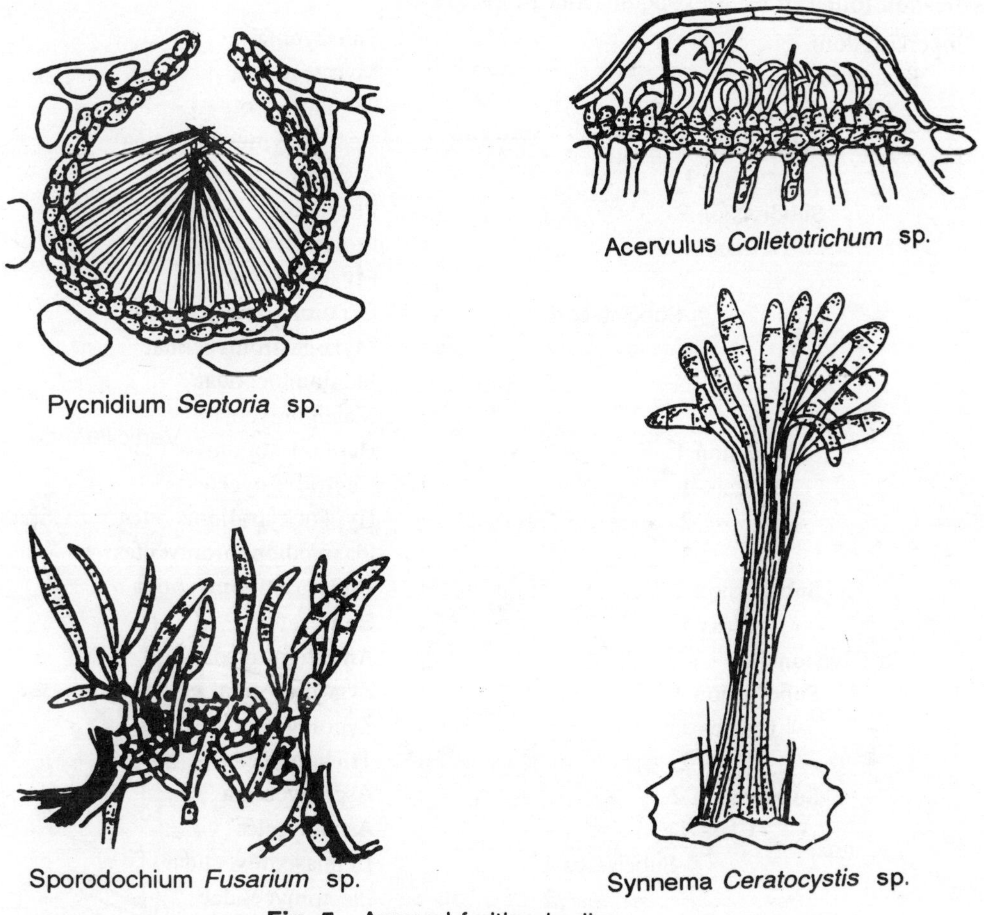

Fig. 5 Asexual fruiting bodies

SEXUAL REPRODUCTION

Sexual reproduction in fungi results in the union of two different nuclei. The process of sexual reproduction consists of three distinct phases.

1. **Plasmogamy:** A union of two protoplasts brings nuclei close together within the same cell.
2. **Karyogamy:** Fusion of the two nuclei occurs.
3. **Meiosis:** The nuclear fusion is followed by meiosis (reduction division). Meiosis reduces the number of chromosomes to haploid.

The sex organs of fungi are called *gametangia.* Gametangia form differentiated sex cells called gametes or may contain one or more gamete nuclei. If gamentangia and gametes are indistinguishable into male and female, they are called *isogamentagia* and *isogametes.* If male and female gamentangia are morphologically different they are called *heterogamentangia* and *heterogametes.* The male gamentagium is called *antheridium* and female gamentagium is called *oogonium.* Motile gametes are called *planogametes.*

CLASSIFICATION OF FUNGI

Fungi have been classified differently by different authors and controversies still persist. The following is the classification followed by Alexopoulos and Mims (1979).

Superkingdom	Eukaryonta
Kingdom	Myceteae
Division I	Gymnomycota
Subdivision 1	Acrasiogymnomycotina
Class 1	Acrasiomycetes
Subdivision 2	Plasmodiogymnomycotina
Class 1	Protosteliomycetes
2	Myxomycetes
Subclasses 1	Ceratiomyxomycetidae
2	Myxogastromycetidae
3	Stemonitocetidae
Division II	Mastigomycota
Subdivision 1	Haplomastigomycotina
Classes 1	Chytridiomycetes
2	Hyphochytridiomycetes
3	Plasmodiophoromycetes
Subdivision 2	Diplomastigomycotina
Class 1	Oomycetes
Division III	Amastigomycota
Subdivision 1	Zygomycotina
Classes 1	Zygomycetes
2	Trichomycetes
Subdivision 2	Ascomycotina
Class 1	Ascomycetes
Subclasses 1	Hemiascomycetidae
2	Plectomycetidae
3	Hymenoascomycetidae
4	Laboulbeniomycetidae
5	Loculoascomycetidae
Subdivision 3	Basidiomycotina
Class 1	Besidiomycetes
Subclasses 1	Holobasidiomycetidae
2	Phragmobasidiomycetidae
3	Teliomycetidae
Subdivision 4	Deuteromycotina
Form - Class 1	Deuteromycetes
Form - Subclasses 1	Blastomycetidae
2	Coelomycetidae
3	Hyphomycetidae

Among the classes and subclasses, some of them do not have any plant pathogen. A few important plant pathogens have been selected from each of the classes and subclasses which contain plant pathogens and their life cycle is described briefly to focus the important structures of these fungi.

CLASS: CHYTRIDIOMYCETES

SYNCHYTRIUM ENDOBIOTICUM

(Order: Chytridiales, Family: Synchytriaceae)

The important character of the class chytridiomycetes is that the zoospore is uniflagellate. The important character of the order Chytridiales is the absence of true mycelium.

The fungus, *Synchytrium endobioticum* causes black wart disease in potato. Uniflagellate zoospores released from infected tissues swim in water present in the soil. The zoospore penetrates inside the epidermis of the host shedding its flagellum outside. Inside the epidermal cell the fungus grows in size and a thick, brown wall is formed. The well-developed structure is called *prosorus.* The host cell also becomes enlarged due to hypertrophy. The surrounding cells also divide (hyperplasia) quickly and tumor or wart-like tissue is formed. The prosorus germinates within the host cell. Repeated mitotic divisions of the nucleus take place. About 32 nuclei are formed. Thin hyaline walls are formed in such a way as to divide the prosorus into four to nine multinucleate segments. Nuclear division continues and about 300 nuclei are formed. Each prosorus portion develops into a sporangium or a gametangium. The mass of sporangia is called the sorus. If water is abundantly present, zoospores are formed. If drought sets in, the motile cells released become planogametes instead of zoospores. They fuse in pairs to form zygotes. The zygote is biflagellate. The zygote penetrates epidermal cell and inside the cell it enlarges and a heavy wall is formed around it. It is now called resting sporangium. From the resting sporangium zoospores are formed. The zoospores are released from the sporangium. They swim and infect plants (Fig. 6).

CLASS : OOMYCETES

PYTHIUM APHANIDERMATUM

(Order : Peronosporales; Family : Pythiaceae)

The important character of the class, Oomycetes is the presence of two flagella of equal size in zoospores. The important character of the order Peronosporales is the eucarpic thallus (forming reproductive structures on certain portions of the thallus). The family Pythiaceae includes fungi bearing their sporangia directly on the somatic hypha. *Pythium aphanidermatum* causes damping-off in many crops like tomato, tobacco and brinjal. The hyphae are coenocytic. Sporangia are formed on the somatic hyphae. Sporangia either germinate directly or they form zoospores. Before the formation of zoospores, a vesicle is formed on a long tube arising form the sporangium. The protoplast of the sporangium moves into the vesicle and differentiation of the zoospores takes place in the vesicle. The vesicular wall bursts and the zoospores swim out. The zoospores are biflagellate. They shed their flagella and encyst. The encysted zoospores germinate by putting forth germ tube.

Sexual reproduction of the fungus is by oogonia and antheridia which are developed in close proximity, often on the same hypha with antheridium just below the oogonium. The oogonium contains a multinucleate oosphere surrounded by a layer of periplasm. The antheridia are small and club-shaped. Antheridium comes into contact with oogonium, a fertilization tube develops and penetrates the oogonial wall and the periplasm. The male nucleus passes through the tube into the oosphere, unites with female nucleus and

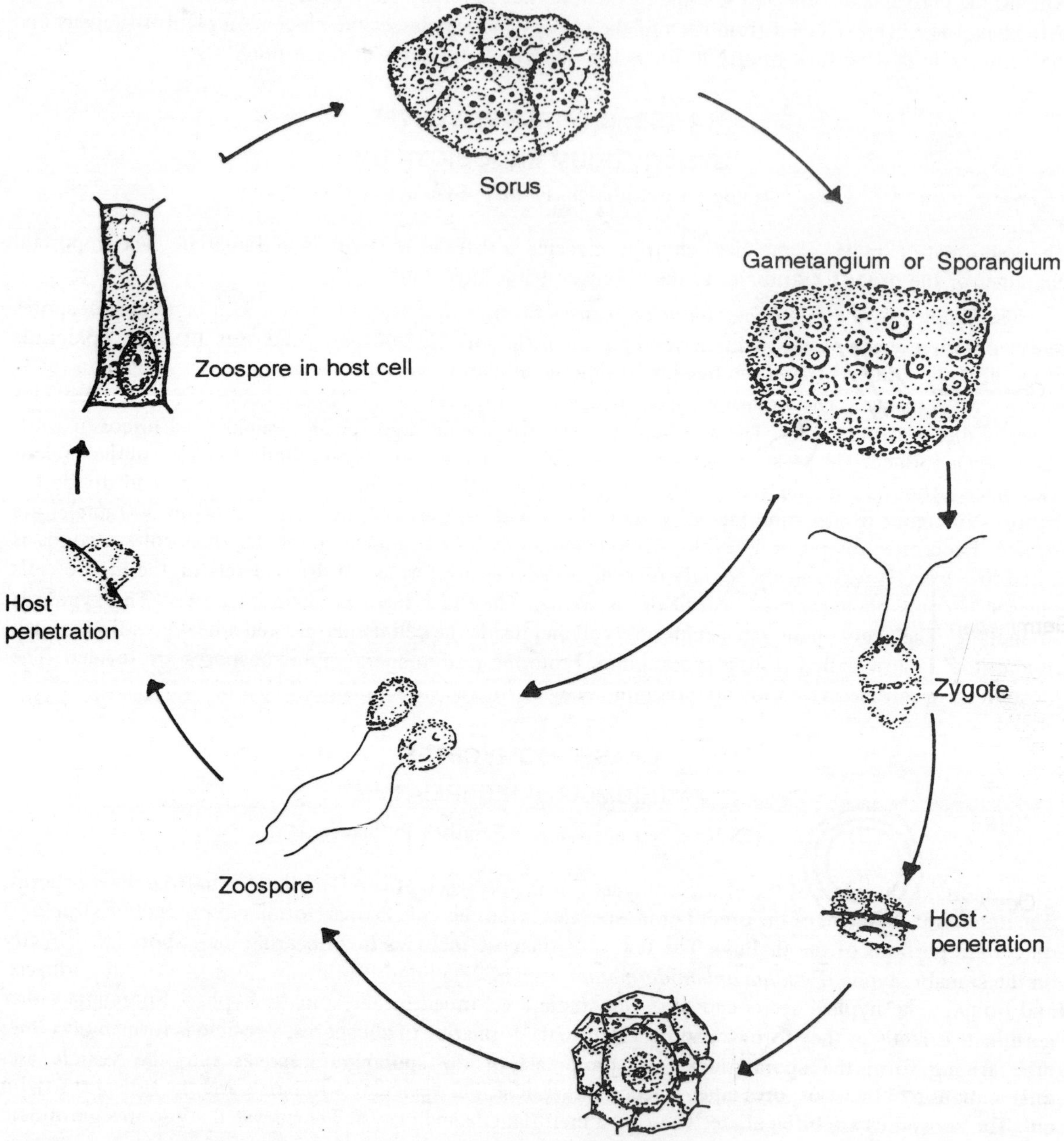

Fig. 6 Key stages in the life cycle of *Synchytrium endobioticum*

forms the zygote. The zygote forms a thick wall and oospore develops. The oospores are the resting spores and they germinate either by producing a germ tube or a vesicle in which zoospores develop (Fig. 7).

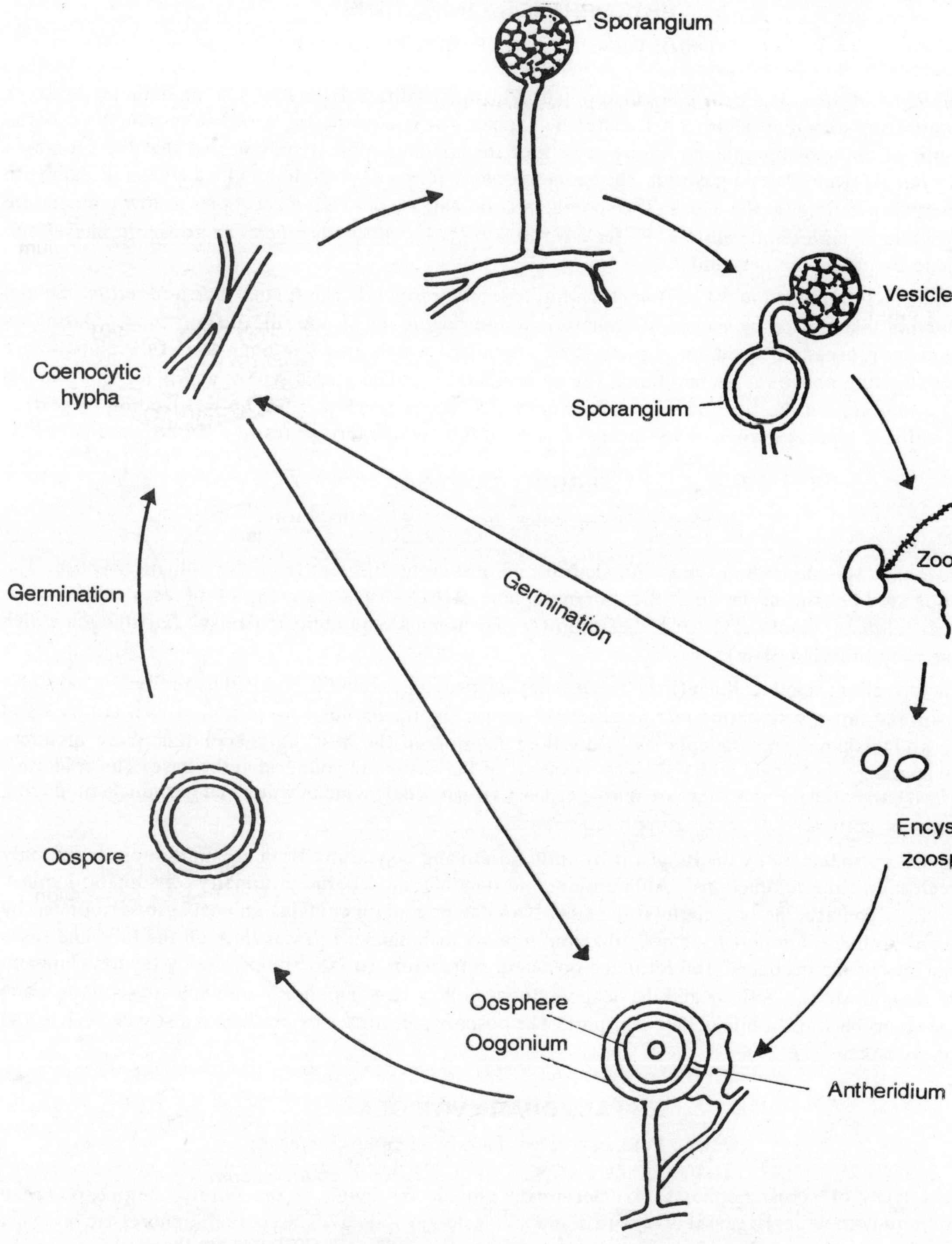

Fig. 7 Key stages in the life cycle of *Pythium aphanidermatum*

PHYTOPHTHORA INFESTANS

(Order: Peronosporales; Family: Pythiaceae)

Phytophthora infestans causes late blight of potato. The fungus differs from *Pythium* sp. in that it produces sporangiophores distinguishable from the somatic hyphae. The lemon-shaped, papillate sporangia are borne at the tips of the sporangiophore. Apical growth of the sporangiophore continues so that the sporangia actually fall off from a lateral position. The sporangiophore forms a sympodium with a zigzag growth with characteristic swellings at the nodes. The sporangia germinate by producing zoospores at low temperature (12°C) while at high temperature (24°) they germinate producing germ tube. The zoospore encyst and germinate by producing germ tube.

For sexual reproduction two different mating types are essential. The fungus is *heterothallic.* Sexual reproduction takes place by means of antheridium and oogonium of opposite mating types. During its development process, antheridium is pierced by oogonium which grows through it and develops into a globose structure above the antheridium. The antheridium is called amphigynous and it forms a funnel-shaped collar around the base of mature oogonium. Oospores develop either by fertilization or parthenocarpically. Oospores germinate by means of a germ tube which terminates in a sporangium (Fig. 8).

ALBUGO CANDIDA

(Order: Peronosporales; Family : Albuginaceae)

In this family sporangiophores are club shaped and strikingly different from the somatic hyphae. The sporangia are in chains at the tip of the sporangiophore. *Albugo* causes a group of diseases called 'white rusts'. *A. candida* attacks cabbage and cauliflower. The fungus is an obligate parasite (a pathogen which can live only in living plant).

The mycelium is intercellular (in between cells) and penetrates into the host cell by producing haustoria (an absorbing organ penetrating into a cell of the host). The intercellular mycelium grows, ramifies and produces club-shaped sporangiophores below the epidermis of the host. Each sporangiophore produces several sporangia in a chain with the oldest at the tip of the chain and youngest at the base. The epidermis of the host bursts and the sporangia are released. They are carried by wind or water and germinate producing zoospores or germ tube depending upon temperature.

Sexual reproduction by the fungus is by antheridium and oogonium. Both are multinucleate but only one nucleus in each is functional. Antheridium and oogonium are borne terminally on somatic hyphae. Antheridium contacts the oogonium at the side. Now the oogonium contains an oosphere surrounded by periplasm. Antheridium puts forth a fertilization tube. A male nucleus passes through the tube and fuses with nucleus of the oosphere. The fertilized oosphere is transformed into an oospore by the development of a thick wall. Meiosis follows and the oospore divides into a large number of uninucleate sections. Each of the sections becomes a biflagellate zoospore. The oospore germinates by producing a sessile vesicle and releases zoospores (Fig. 9).

PLASMOPARA VITICOLA

(Order : Peronosporales; Family : Peronosporaceae)

Various types of sporangiophores of determinate growth are found in the family, Peronosporaceae. *Plasmopara viticola* is the causal organism of grapevine downy mildew. The sporangiophores are branched ones and sporangia are borne in clusters on sterigmata at the tips of the branches. Sporangia germinate by producing biflagallate zoospores. The zoospores encyst and then germinate. Oogonium and antheridium are

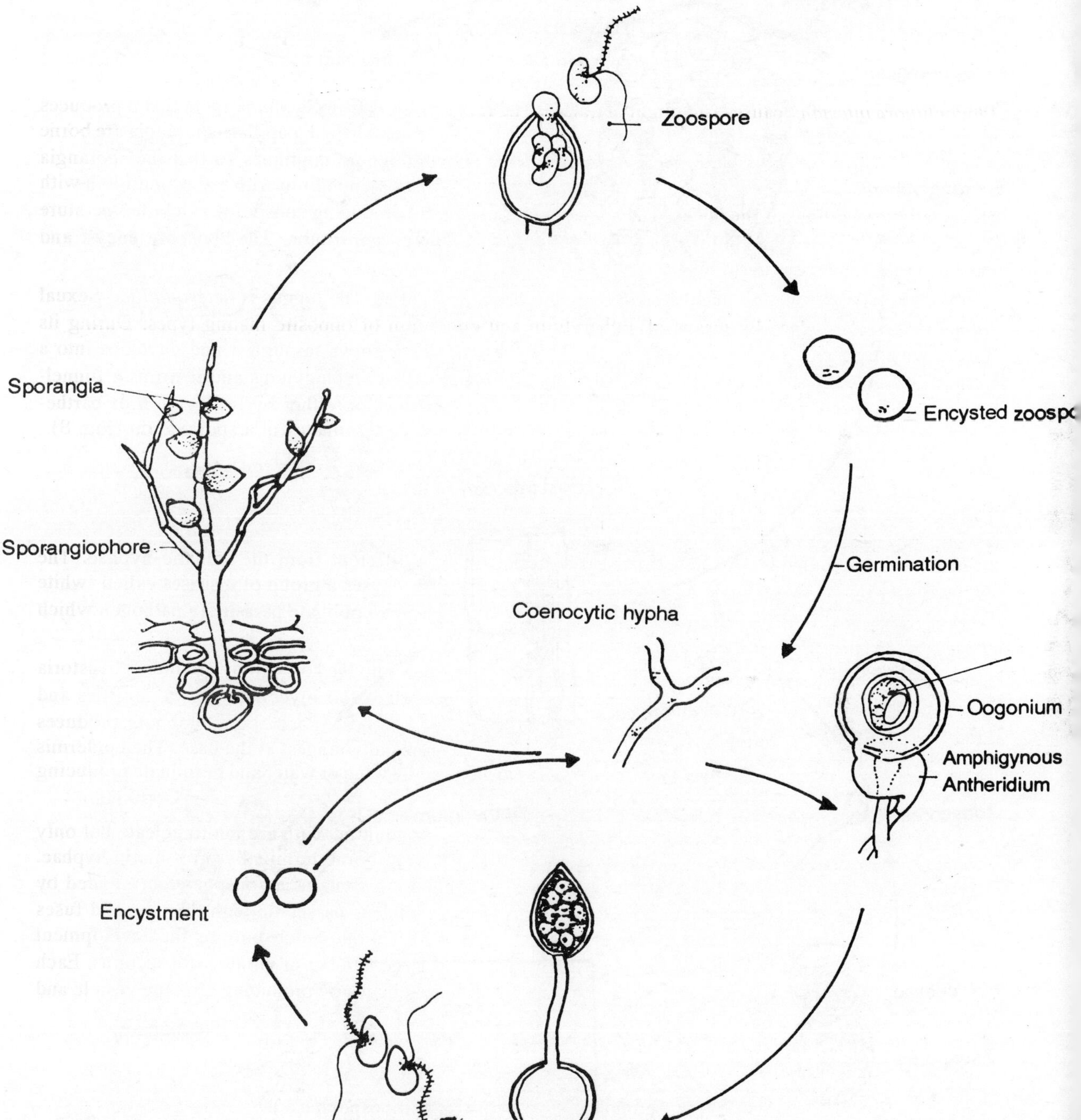

Fig. 8 Key stages in the life cycle of *Phytophthora infestans*

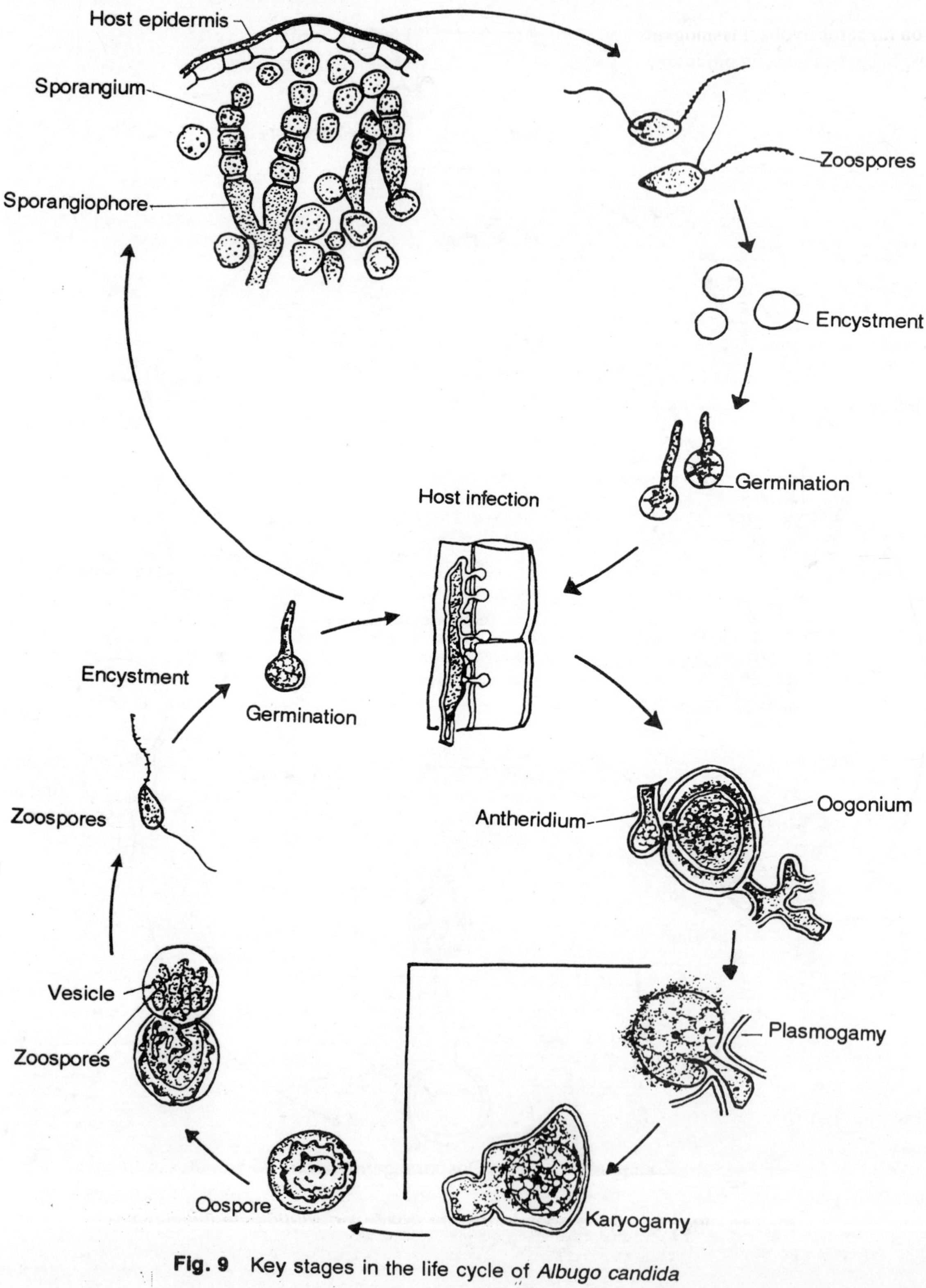

Fig. 9 Key stages in the life cycle of *Albugo candida*

found on the same hypha. Plasmogamy and karyogamy lead to the formation of oospore. Meiosis takes place and oospore germinates by producing a sporangium. From the sporangium zoospores are released (Fig. 10).

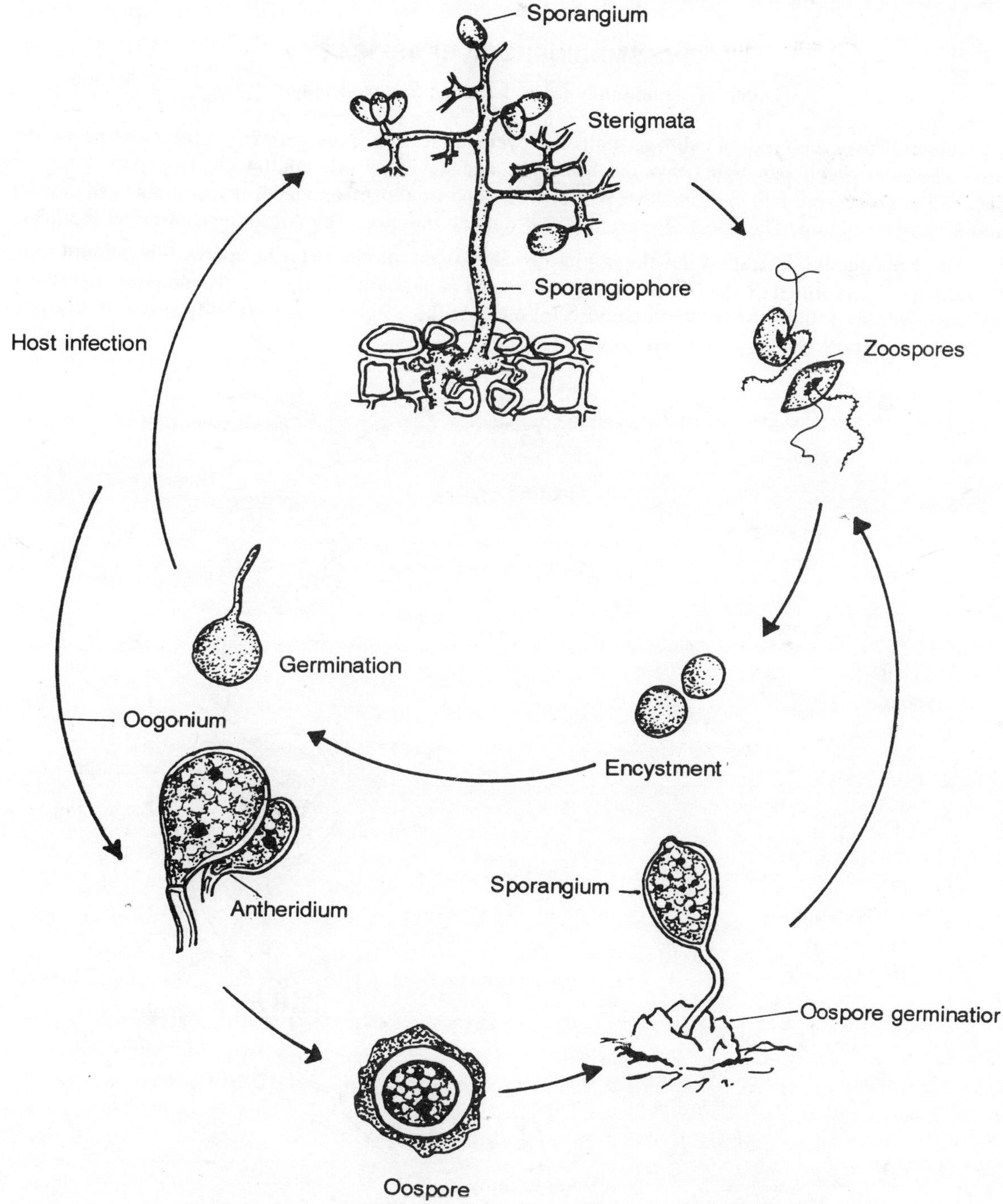

Fig. 10 **Key stages in the life cycle of** ***Plasmopara viticola***

CLASS : PLASMODIOPHOROMYCETES

The important character of the class is the presence of plasmodium (a naked mass protoplasm with many nuclei, moving and feeding in amoeboid fashion).

PLASMODIOPHORA BRASSICAE

(Order : Plasmodiophorales; Family : Plasmodiophoraceae)

The fungus causes club root of cabbage. In its life cycle the resting spore germinates by releasing a swarm cell. The swarm cell penetrates into the host cells and develops into multinucleate plasmodium. The plasmodium is cleaved into multinucleate portions. Each portion is surrounded by a membrane and develops into a zoosporangium. The zoospores are released outside the host. The zoospores can infect the host.

Some plasmodia, instead of developing into zoosporangia, produce resting spores. The resting spores would have been formed by the formation of gametes and copulation of gametes. Plasmogamy, karyogamy and meiosis may follow. Akaryote phase also follows. At this stage the nuclear body seems to disappear and after this stage resting spores are formed (Fig. 11).

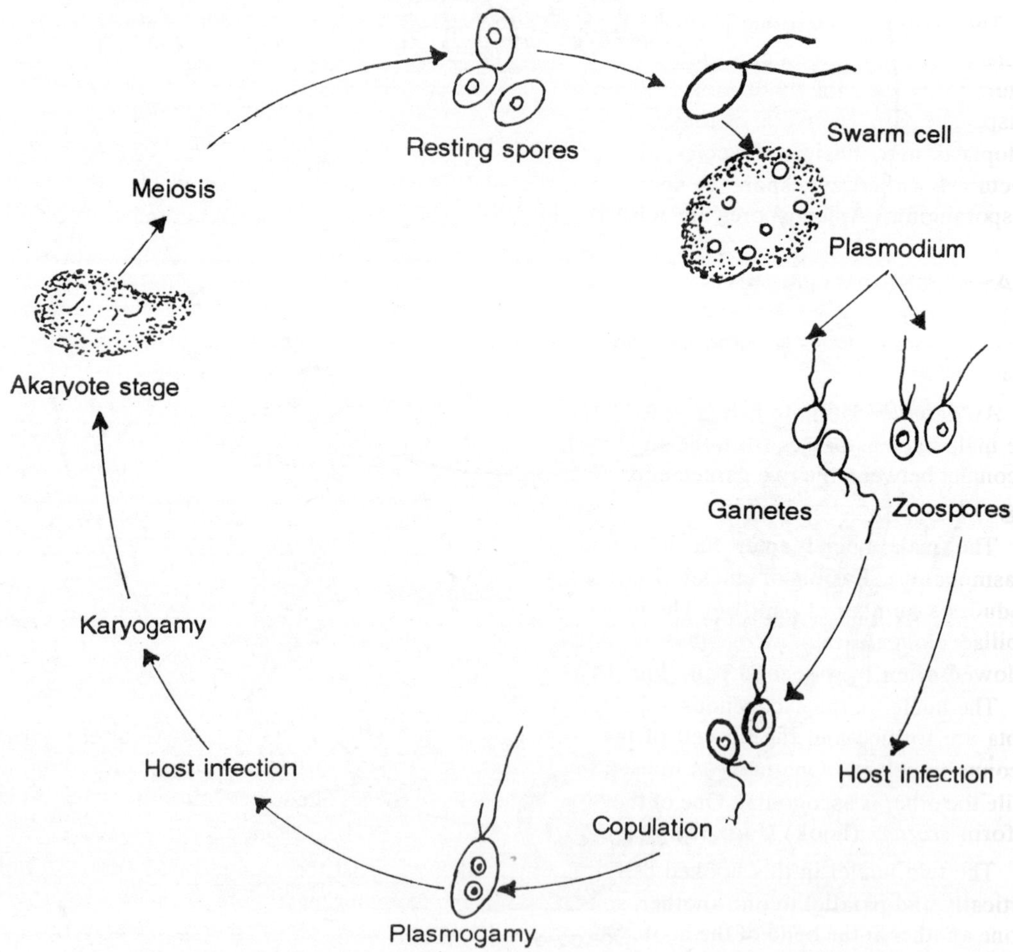

Fig. 11 Key stages in the life cycle of *Plasmodiophora brassicae*

CLASS : ZYGOMYCETES

The production of zygospore is characteristic of this class. Zygospore results from the complete fusion of two gametangia while oospore found in other fungi arises from an oosphere found inside the oogonium. Zoospores (motile spores) are absent in this class and the spores are aplanate (non-motile).

RHIZOPUS NIGRICANS

(Order : Mucorales: Family : Mucoraceae)

It causes fruit rot of jack and soft rot of sweet potato. The mycelium is coenocytic (non-septate). It develops rhizoids (a short branch of a thallus which looks like a root) at certain points. Directly above the rhizoids, sporangiophores are produced. The top of the sporangiophores becomes swollen and a sporangium develops. The central portion of the sporangium is a columella (a sterile structure which is extension of the stalk). The peripheral zone is the spore-bearing portion of the sporangium. Sporangiospores are produced inside the sporangium and they are aplanospores.

The fungus is heterothallic (two different thalli required for sexual reproduction). When two opposite strains (+ and -) come in contact with one another, progamentangia (before gametangium) are formed. A septum forms near the tip of each progametangium, separating into two cells, a terminal gametangium and a suspensor cell. The walls of the two contacting gametangia dissolve at the point of contact and the two protoplasts mix. Fusion of nuclei takes place. The cell wall of the fused gametangia thickens and the structure is called zygospore. A sporangiophore arises and develops into a sporangium which is called zoosporangium. Aplanospores are released (Fig. 12).

CLASS : ASCOMYCETES

The class Ascomycetes is characterized by septate mycelium and by the formation of ascogonium, ascocarp, ascus, and ascospores.

Ascogonium is the female gametangium of the Ascomycetes. The male gametangium is antheridium. The male nucleus passes from the antheridium into the ascogonium through a pore developed at the point of contact between the two gametangia. Ascogonium contains *trichogyne* which receives the male nucleus (Fig. 13).

The male nuclei enter the ascogonium and in many cases they pair with ascogonial nucleus (plasmogamy). Fusion of nuclei (karyogamy) does not take place immediately. Now the ascogonium produces a number of papillae. The nuclei from the ascogonium begin to pass into them one by one. The papillae elongate into *ascogenous hyphae.* In this ascogenous hyphae leading pair of nuclei appear, followed often by a second pair (Fig. 14).

The nuclei in the ascogenous hyphae and those still in the ascogonium undergo simultaneous mitosis. Septa are formed and the tip cell of the ascogenous hypha becomes uninucleate and several other cells become binucleate. One nucleus in each binucleate cell of the ascogenous hyphae is antheridial in origin while the other is ascogonial. One of the binucleate cells of the ascogenous hypha elongates and bends over to form *crozier* (hook) (Fig. 15).

The two nuclei in this hooked cell divide in such a way that their spindles are oriented more or less vertically and parallel to one another, so that two of the daughter nuclei, one from each spindle, are close to one another at the bend of the hook. One of the other two nuclei is located at the tip and the other nucleus is near the basal septum of the hook. Thus, the crook cell is binucleate while the tip and basal cells are uniculeate.

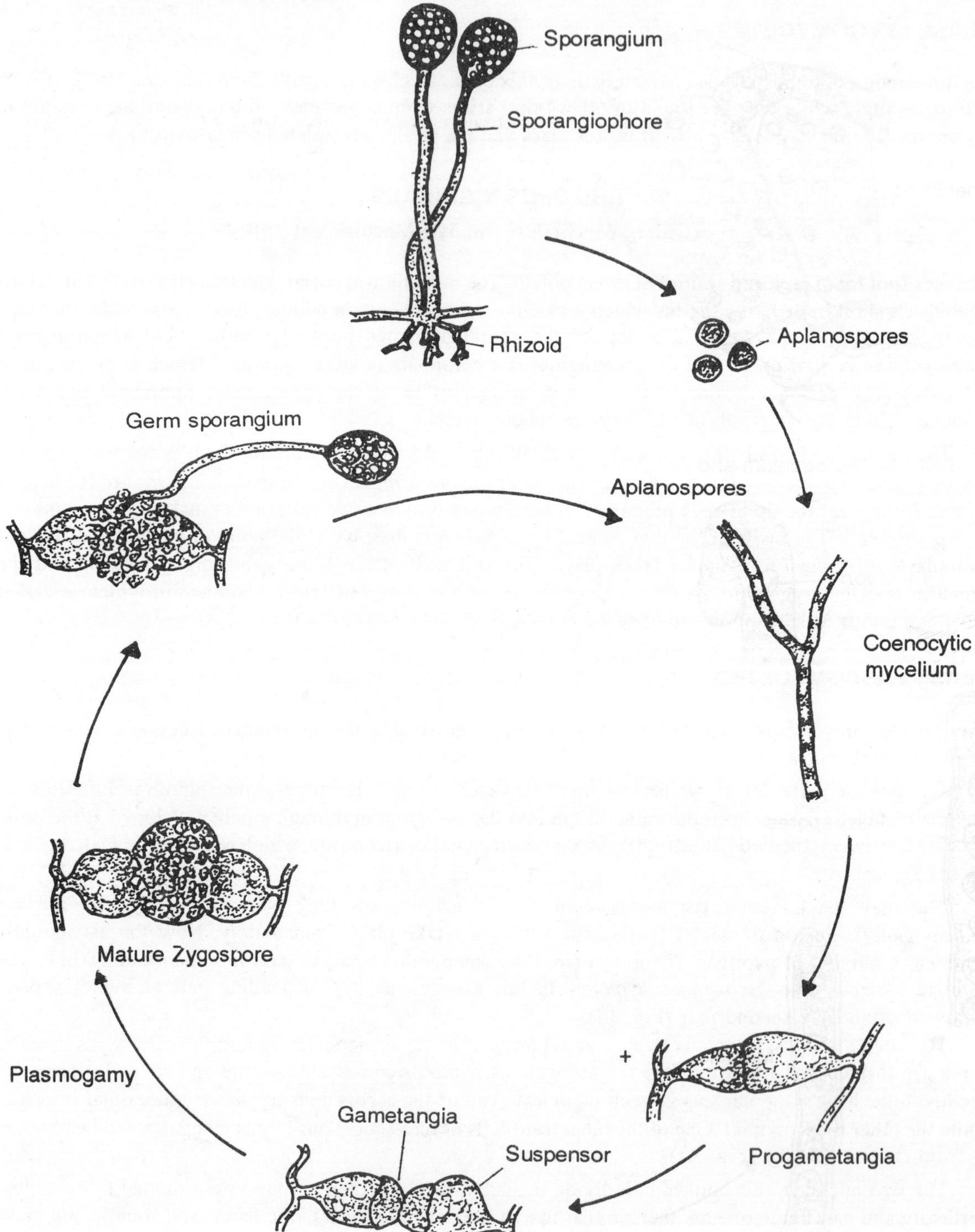

Fig. 12 Key stages in the life cycle of *Rhizopus nigricans*

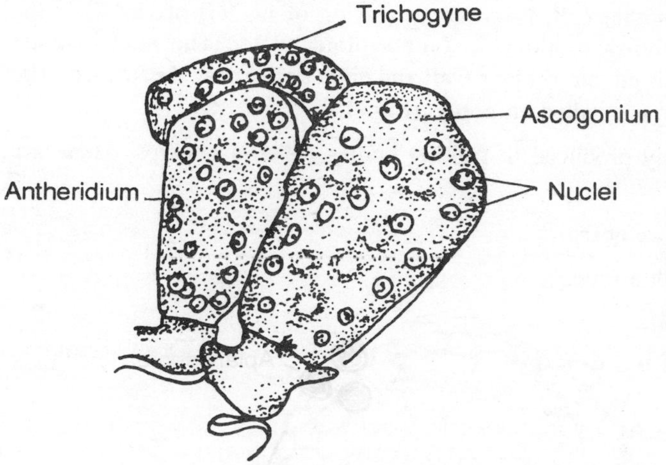

Fig. 13 Ascogonium and Antheridium

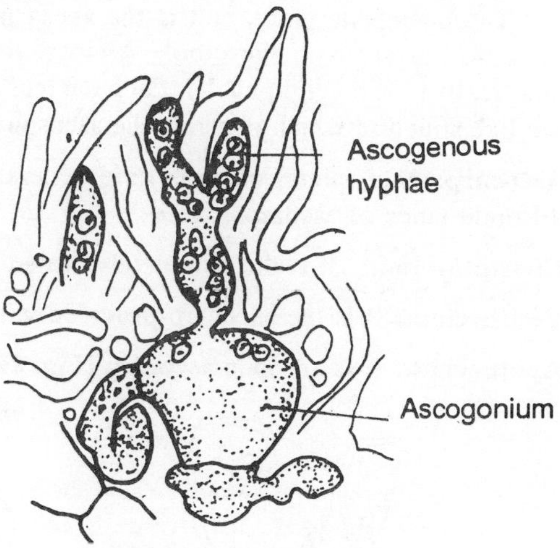

Fig. 14 Ascogenous hyphae

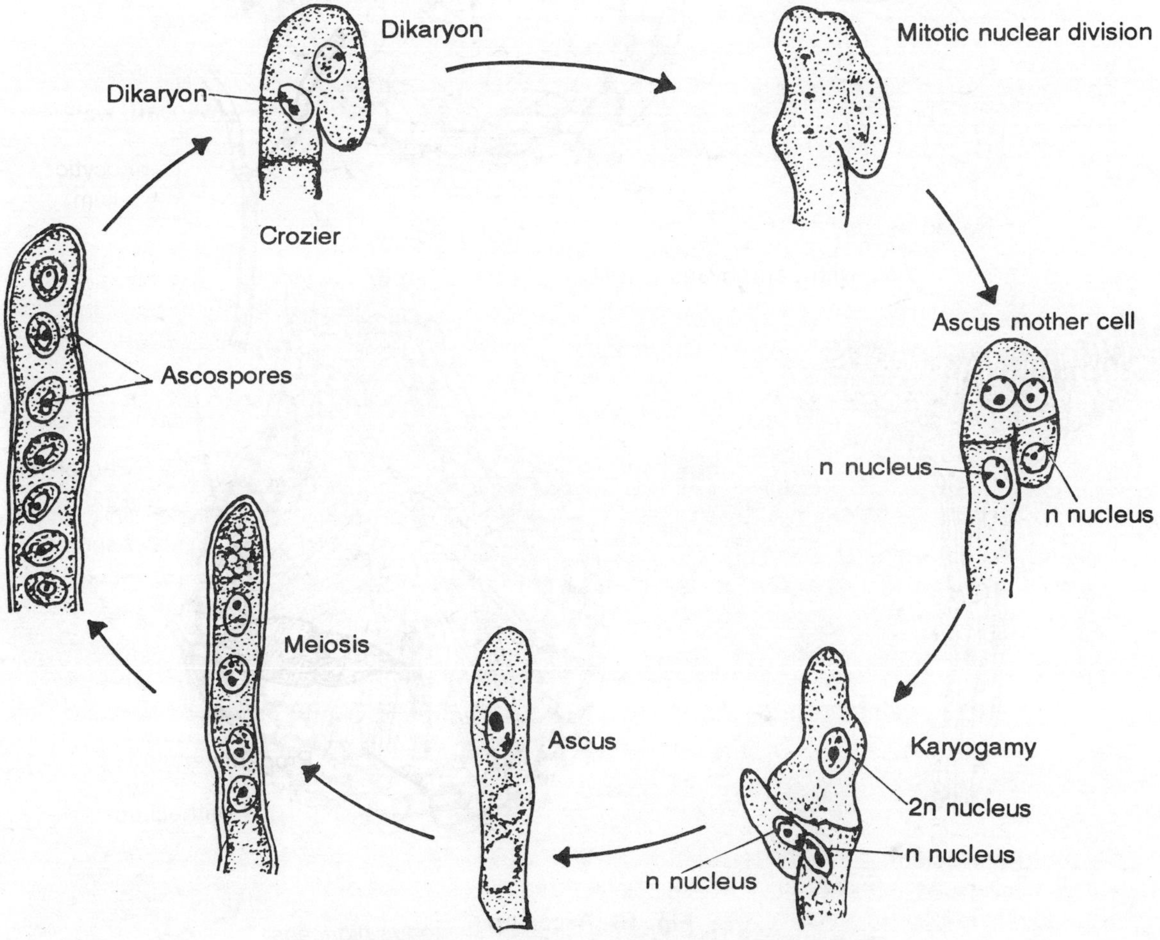

Fig. 15 Formation of ascus and ascospores

The binucleate-crook cell is the ascus mother cell. Karyogamy (fusion of nuclei) occurs after the formation of septa in the crook. Meiosis follows resulting in four haploid nuclei. The nuclei divide mitotically to form eight nuclei. Each nucleus is enveloped by a wall and eight *ascospores* are formed. The sac-like structure which contains the ascospores is called an *ascus* (Fig. 15).

Ascocarp: Asci (plural of ascus) are generally produced in fruiting bodies called ascocarps. There are different kinds of ascoarps.

Cleistothecium: It is the completely closed ascocarp.

Perithecium: It is the ascocarp provided with a pore.

Apothecium: It is an open ascocarp (Fig. 16).

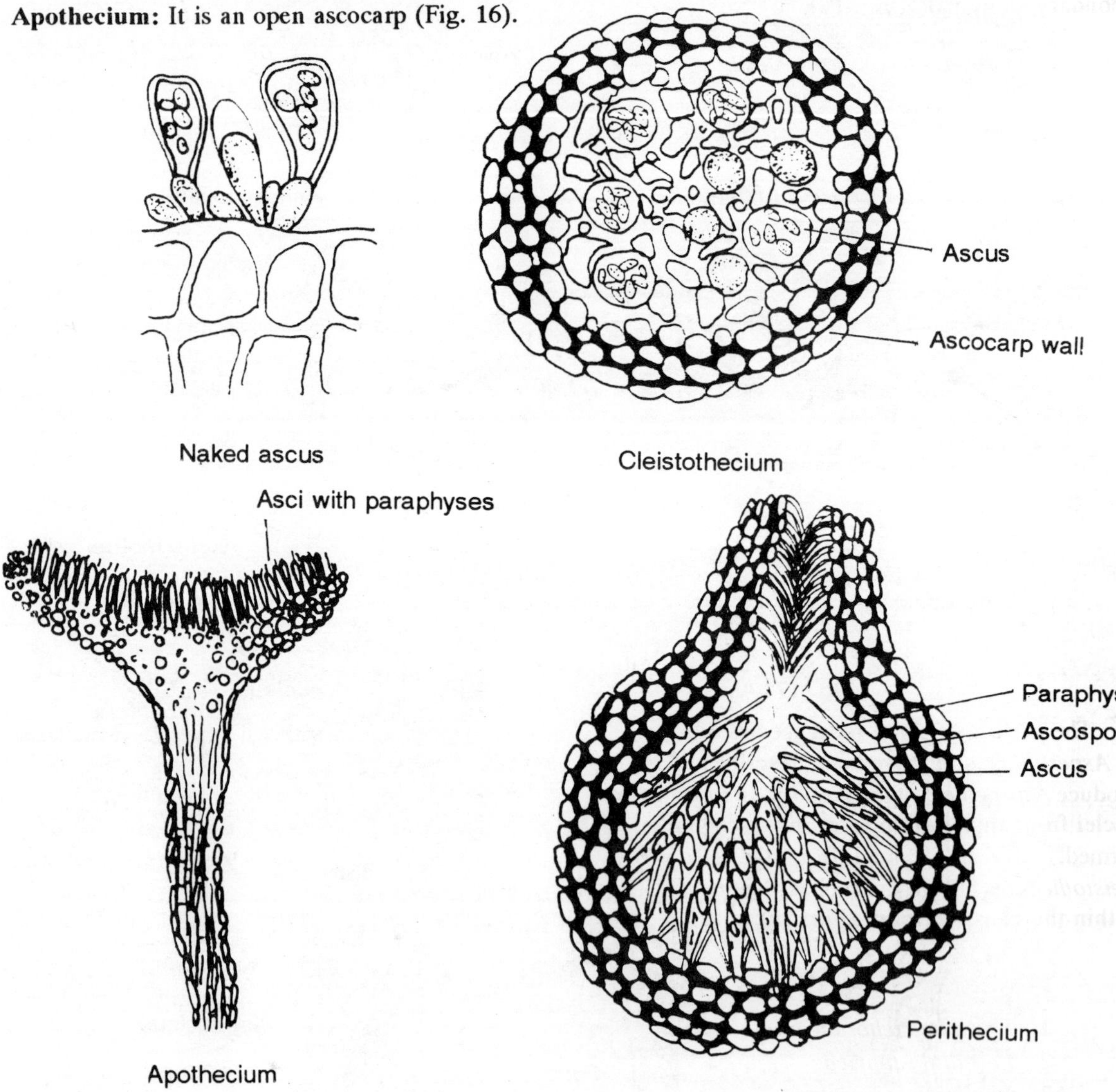

Fig. 16 Ascocarp

ASPERGILLUS NIGER

(Subclass : Plectomycetidae; Order : Eurotiales; Family : Eurotiaceae)

It is a common saprophyte (utilizes dead organic matter for food). It also causes jack fruit rot and sweet potato tuber rot. The hyphae are septate and multinucleate. The mycelium produces conidiophores. The hyphal cell which gives rise to the condiophore is called *foot cell.* The conidiophores terminate in a bulbous head, the *vesicle.* A large number of sterigmata are produced over the surface of the vesicle. One or two layers of sterigmata are produced. The first layer is called primary *sterigmata* and the upper most layer which is typically bottle shaped are called secondary sterigmata. Conidia are formed in chains on the secondary sterigmata (Fig. 17).

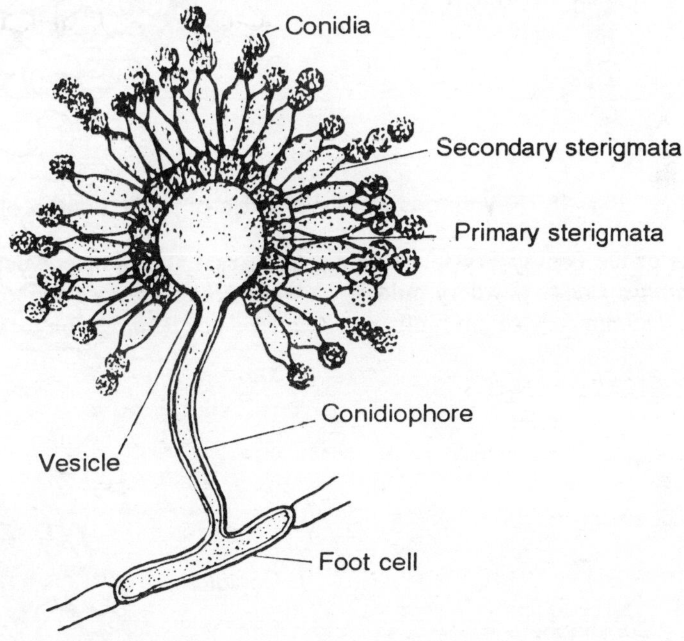

Fig. 17 Asexual reproduction in *Aspergillus niger*

The perfect stages of most species of *Aspergillus* species have not been detected. Three perfect stages of *Aspergillus* spp. have been described. They are *Eurotium, Sartorya* and *Emericella* and all of them produce *Aspergillus* - type conidia. In *Eurotium* antheridia and ascogonia coil around each other. Probably nuclei from antheridium enter into ascogonium and pairing of nuclei takes place. Ascogenous hyphae are formed. Asci are formed from the ascogenous hyphae and asci and enclosed in closed ascocarp, *cleistothecium.* From the asci ascospores are formed and asci disappear soon leaving the ascospores free within the cleistothecium (Fig. 18).

POWDERY MILDEWS

1. *Erysiphe cichoracearum*
2. *Leveillula taurica*
3. *Phyllactinia corylea*

(Subclass : Hymenoascomycetidae; Series : Pyrenomycetes; Order : Erysiphales; Family : Erysiphaceae)

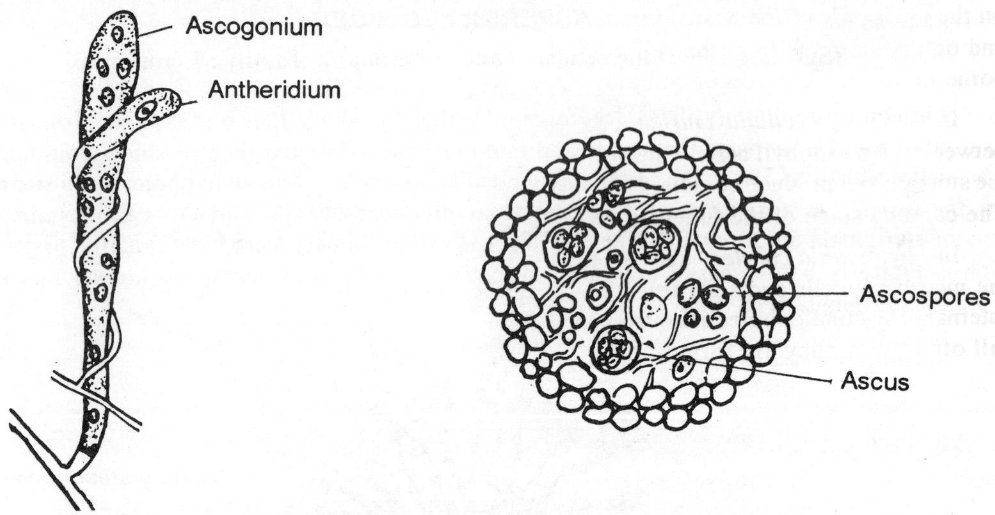

Fig. 18 Key stages in sexual cycle of perfect stage of *Aspergillus*

The members of the Family Erysiphaceae are the causal agents of powdery mildews in different crops. *Erysiphe graminis* causes powdery mildew on wheat, *E. cichoracearum* on cucurbits, *Leveillula taurica* on chilli, *Phyllactinia corylea* on mulberry, *Sphaerotheca pannosa* on rose and *Podosphaera leucotricha* on apple.

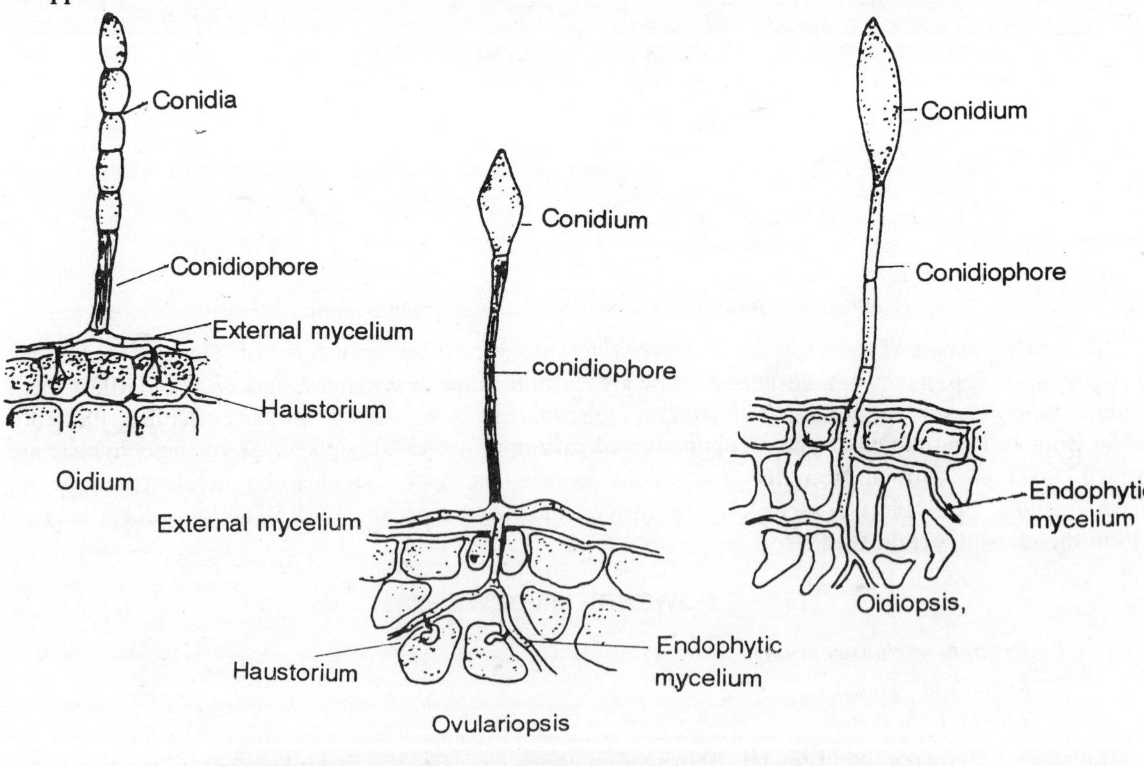

Fig. 19 Asexual stages of three types of powdery mildew fungi

All the powdery mildew fungi are obligate parasites. The mycelium of *E. cichoracearum* is superficial on the epidermis of the host. From the somatic hypha, haustoria arise, penetrate into the epidermal cells and obtain nutrients from the cell sap. From the surface mycelium, conidiophores arise and condidia are borne in chains at the tip of conidiophore. The conidial stage of the fungus is called *'Oidium'*

In case of *Leveillula taurica,* the hyphae penetrate into the leaf through the stomata, and spread between the mesophyll cells. Thus the mycelium is internal or endophytic. The conidiophores grow out of the stomata and produce conidia. No conidial chains are formed; conidia fall off as soon as they are formed. The conidial stage of the fungus is called *'Oidiopsis'.*

In *Phyllactinia corylea,* most of the hyphae are superficial but some hyphae enter the stomata, reach the mesophyll cells and send haustoria into them. Thus the mycelium of the fungus is both external and internal. The condiophores arise from external mycelium and produce single clavate shaped conidia which fall off as soon they are formed. The conidial stage of the fungus is called *'Ovulariopsis'* (Fig. 19).

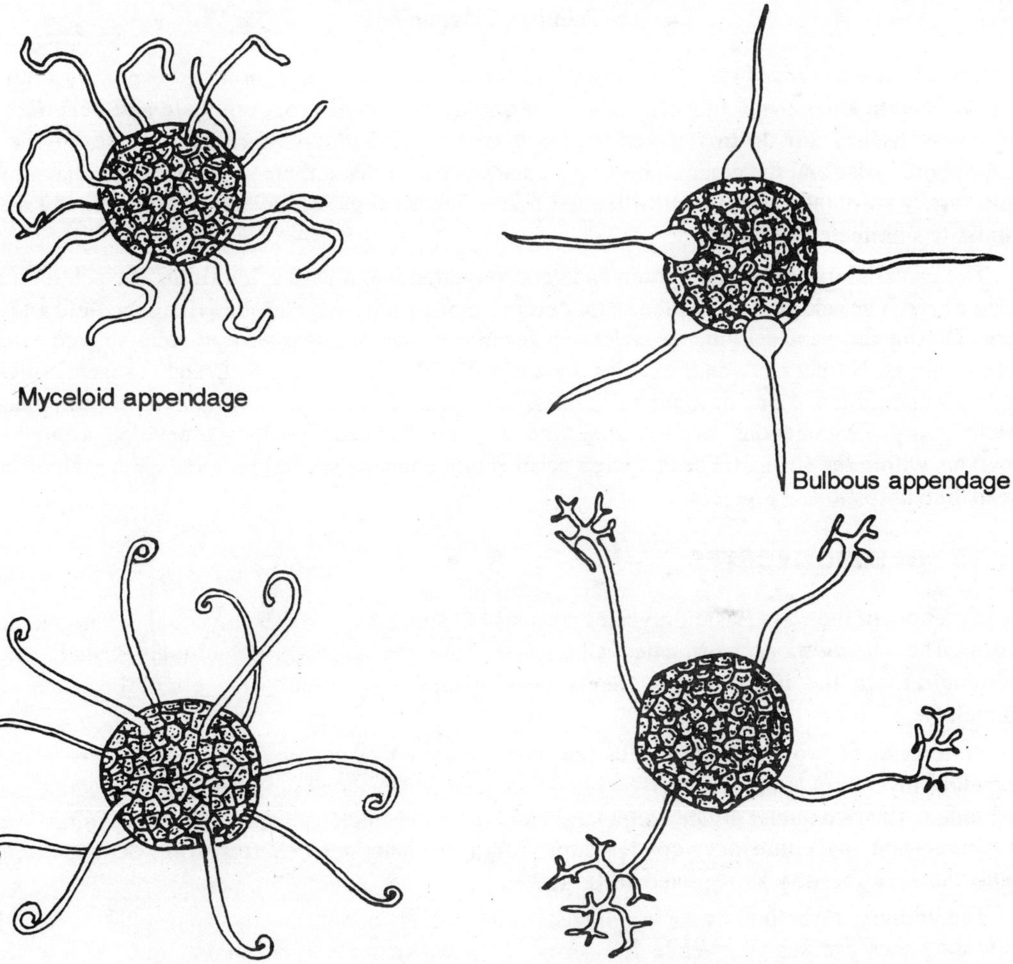

Fig. 20 Cleistothecia of powdery mildew fungi

The sexual stages appear in the later stages of their life cycle. The antheridium is slender than the ascogonium. They press closely together. The antheridial nucleus passes into the ascogonium through a pore. Ascogenous hyphae, asci and ascospores are formed. The ascocarp is a cleistothecium. The cleistothecia have appendages. Myceloid appendages are found in cleistothecia of *Erysiphe* and *Leveillula.* Bulbous appendages are seen in *Phyllactinia.* Cleistothecia of *Uncinula* has appendages with rigid curled tips (circinoid) while dischotomously branched appendages are seen in cleistothecia of *Podosphaera* (Fig. 20).

Many asci are found inside the cleistothecia of *Erysiphe, Leveillula* and *Phyllactinia. Sphaerotheca* and *Podosphaera* produce only one large ascus in each cleistothecium. Number of ascospores inside the ascus may vary from two to eight

CLAVICEPS PURPUREA

Subclass : Hymenoascomycetidae; Series : Pyrenomycetes; Order : Clavicipitales; Family : Clavicipitaceae)

The fungus causes ergot of rye. The ascocarp in the fungus is a perithecium. Ascospores are released from the perithecium and carried by wind. They fall on the flowers of rye, germinate and penetrate the ovary. The ovary tissues are destroyed and replaced by white, cottony mycelial mat. From the mat short condiophores arise and the conidiophores produce small conidia at their tips. These conidia are mixed with some sugary substance. The insects attracted by the sugar exudation invade the flowers and distribute the conidia to uninfected flowers.

The mycelial mat begins to harden and gets converted into a hard sclerotium. The sclerotia replace the grains of rye. The sclerotium is called as the 'ergot'. During harvest sclerotia fall on the field and overwinter there. During the next season, the sclerotia germinate and produce several long-stalked stromata with globose heads. Number of small cavities are formed within the stromatal heads. These cavities contain a single ascogonium and one or more antheridia. The male nuclei migrate from antheridium to ascogonium (plasmogamy). Ascogenous hyphae arise and asci are formed. Perithecia develop around the sexual apparatus within the stromatal heads. Each perithecium contains several asci and each ascus contains eight thread-like ascospores (Fig. 21).

CLASS : BASIDIOMYCETES

The mycelium of the class Basidiomycetes consists of septate hyphae. The mycelial development is in three stages. The *basidiospore* germinates and grows into the primary mycelium. Primary mycelium is multinucleate in the initial stages. Septa develop and the primary mycelium becomes septate and uninucleate.

Protoplasts of two uninucleate cells fuse without actual karyogamy and binucleate cells develop. The secondary mycelium is binucleate. The binucleate cell produces a branch and the two nuclei migrate into the branch. The two nuclei divide conjugately and the sister nuclei separate into two daughter cells. In this way binucleate mycelium develops. In some fungi the daughter cell formation occurs through *'clamp connections'* in the way as depicted in (Fig. 22).

The tertiary mycelium is an organized tissue which composes the sporophores. The hyphae of Basidiomycetes are septate. While the septa of Ascomycetes are perforated in a simple manner, the Basidiomycetes contain *dolipore* septum. The dolipore septum looks like a barrel with open ends.

The sexual structure of Basidiomycetes is the *basidium.* The basidium originates as a terminal cell of a binucleate hypha. The basidium becomes broader. The two nuclei fuse (karyogamy). The zygote nucleus

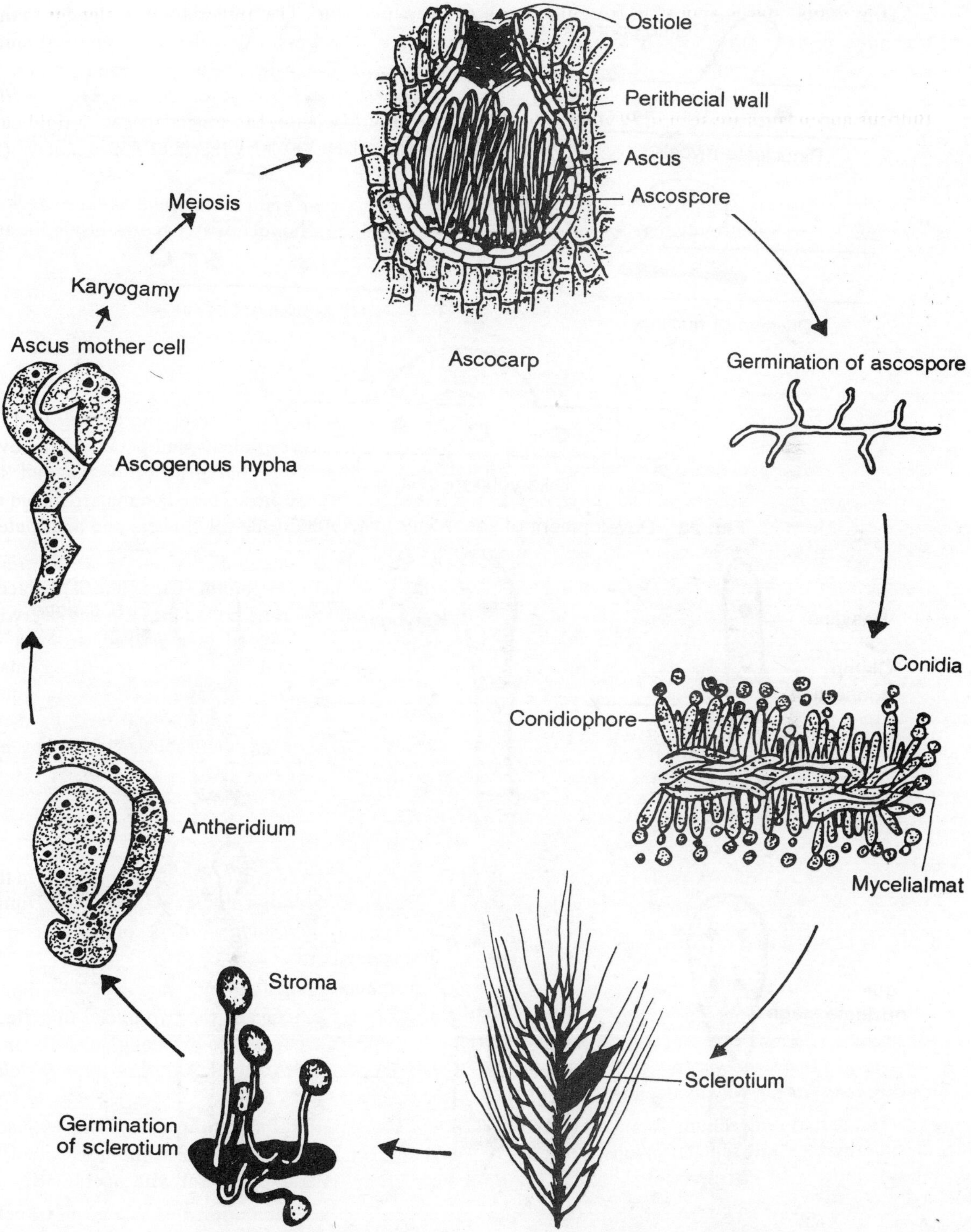

Fig. 21 Key stages in the life cycle of *Claviceps* purpurea

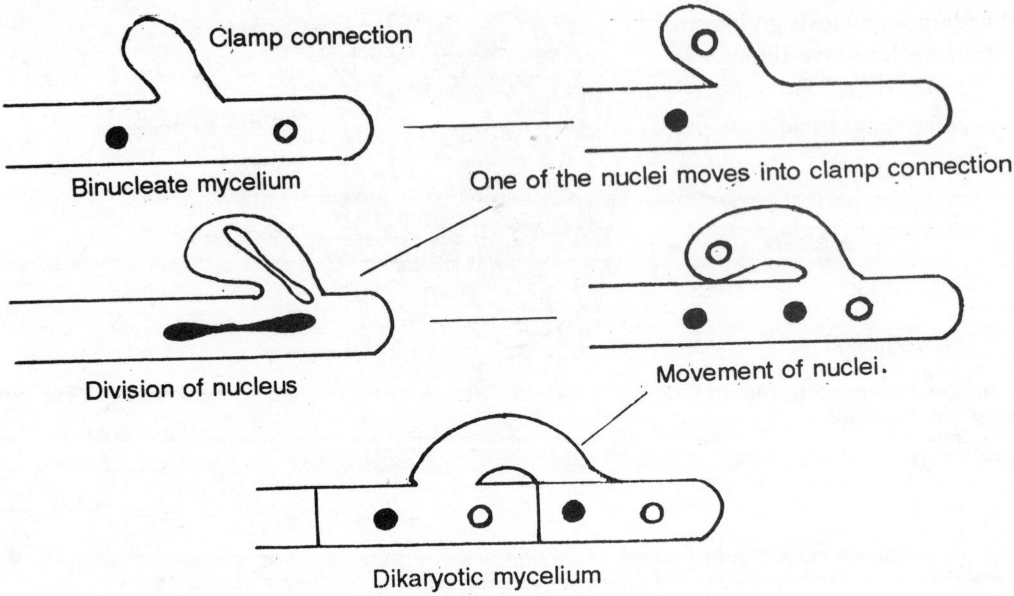

Fig. 22 Development of secondary (dikaryotic) mycelium

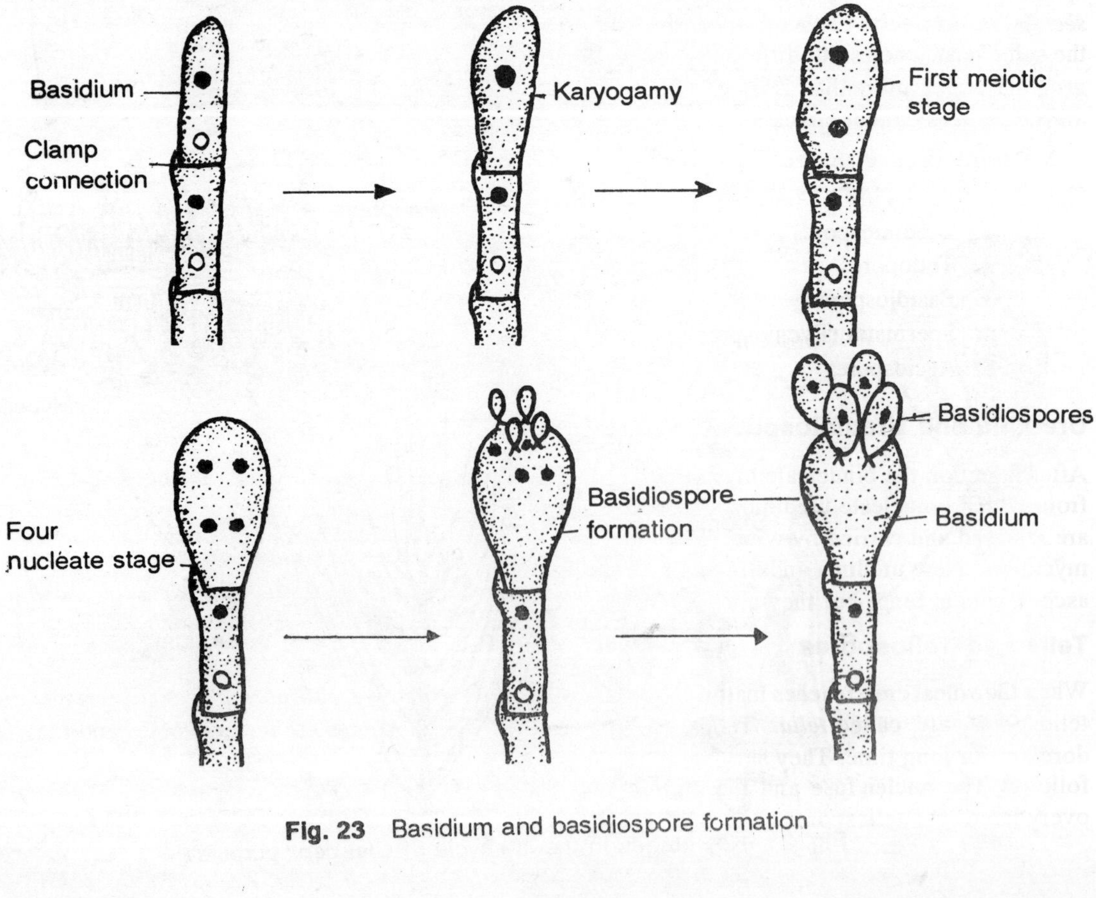

Fig. 23 Basidium and basidiospore formation

(2n) undergoes meiosis giving rise to four haploid nuclei. Four *sterigmata* arise at the top of the basidium. The four nuclei move through sterigmata into young basidiospores. Uninucleate basidiospores develop subsequently (Fig. 23).

In some fungi basidia are produced in highly organized fruiting bodies called *basidiocarp* similar to ascocarp.

Thus, the important characters of Basidiomycetes are: basidium, dikaryotic mycelium, clamp connections and dolipore septum.

PUCCINIA GRAMINIS TRITICI

(Sub class : Teliomycetidae; Order : Uredinales; Family : Pucciniaceae)

The fungus causes stem rust in wheat. It is an obligate parasite. It is a heteroecious rust. The *heteroecious* rust requires two distinct host species to complete its life cycle. It is against *autoecious* rust which completes its entire life cycle on a single host species. Example for autoecious rust is *Melampsora lini.*

The fungal name is having three names (trinomical):

Puccinia graminis tritici: The last name indicates the name of subspecies. A number of biological subspecies exist in *P. graminis.* The sub-species are morphologically similar but differ only in their ability to attack host species. The sub-species which attacks wheat is called *P. graminis tritici.* while the subspecies attacking barley, oats and rye are called *P. graminis hordei, P. graminis avenae* and *P. graminis secalis* respectively. Each sub-species is composed of a large number of physiological races. Races attack the same host species but differ in their reaction to different cultivars of the same host. For example in *P. graminis tritici* more than 200 races differing in their ability to attack various varieties of wheat exist and all those races cannot attack any crop other than wheat.

The life cycle of *P. graminis tritici* is completed in wheat and barberry bush (a wild plant). In its life cycle the fungus produces five different spores.

1. Urediniospores
2. Teliopores
3. Basidiospores
4. Spermatia (Pycniospores)
5. Aeciospores

Uredinia and Urediniospores

After infection the binucleate mycelium develops in wheat plants. It forms masses of cells called *uredinia* from which binucleate urediniospores arise on long stalks. The host epidermis breaks and urediniospores are released and carried by wind. When they fall on wheat leaves, they germinate and produce binucleate mycelium. New uredinia and urediniospores are produced. Urediniospores can be compared to conidia of ascomycetous fungi but they are binucleate (Fig. 24).

Telia and Teliospores

When the wheat crop reaches maturity, the uredinia begin to produce *teliospores.* The pustules that produce teliospores are called *telia.* Teliospores are two-celled. Teliospores are resting spores and they remain dormant for long time. They survive on the stubbles. Initially teliospores are binucleate, soon karyogamy follows. The nuclei fuse and the diploid (2n) nuclei are formed. The teliospores in the diploid stage overwinter. The teliospores cannot infect wheat crop.

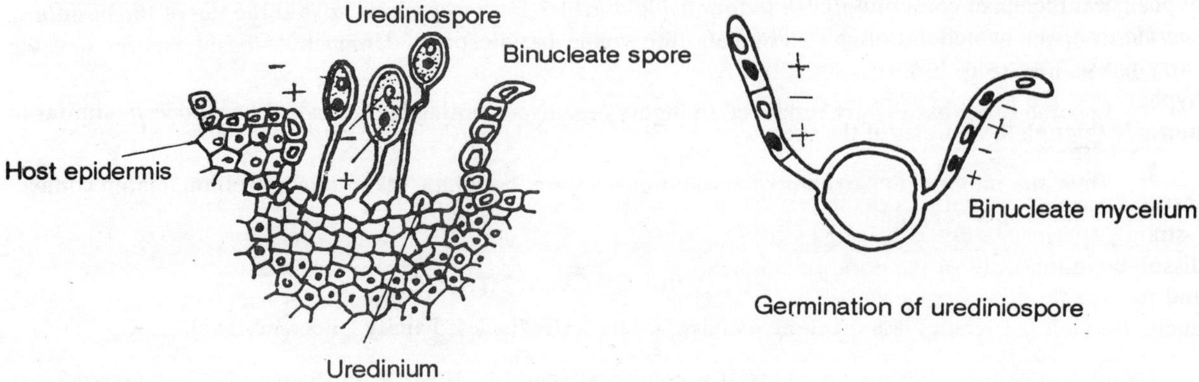

Fig. 24 Uredinia and Urediniospore

Promycelium and Basidiospores

After a resting period, the teliospores germinate producing a *promycelium* (or metabasidium). The diploid nucleus moves into the promycelium and meiosis takes place resulting in four haploid nuclei. Septa are formed separating the nuclei from one another into four cells. A *sterigma* is produced from each cell of the promycelium. A minute *basidiospore* is formed on the sterigma. The nuclei squeeze through the sterigmata into the basidiospores. Two of the nuclei will be of one strain and the other two will be of the other strain (Fig. 25). Basidiospores are released and carried by wind. The basidiospores cannot infect wheat plants. They can infect only barberry bush.

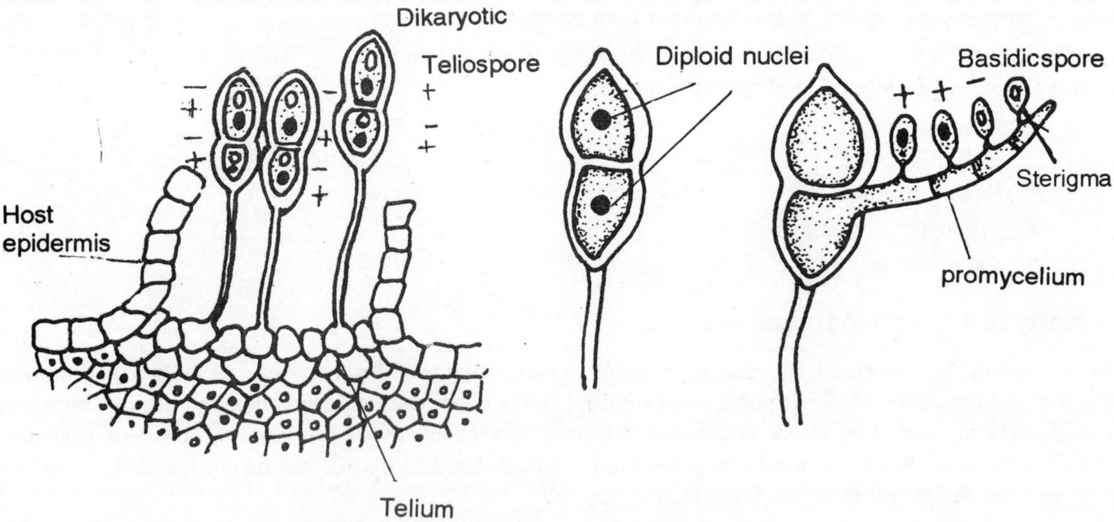

Fig. 25 Formation of basidiospores

Spermogonia (Pycnia) and Spermatia (Pycniospores)

The basidiospores germinate and penetrate the barberry leaves. The uninucleate mycleium multiplies inside the host intercellularly and produces haustoria which absorb nutrients from the cells. A few days later the hyphae near the upper epidermis develop *spermogonium (pycnium)*. The spermogonium contains numerous *spermatiophores* which cut off a succession of minute *spermatia.* Each spermatium contains a nucleus carrying the + or — factor. All spermatia from a single spermogonium carry the same factor. Receptive hyphae with the same genetic make-up as the spermatia also arise along with the spermatiophore. They protrude through the ostioles of the spermogonia. Several periphyses (hair-like sterile structures) also arise.

Insects are attracted towards spermogonia because of sweet nectar exuding from the ostioles. If the insects transmit spermatia from one spermogonium (+ strain) to receptive hyphae of another spermogonium (-strain), spermatization is effected. The spermatial contents pass into the receptive hyphae by a pore dissolved in the walls at the point of contact. During this period, the mycelium penetrates the entire leaf and reaches the lower epidermis. Near the lower epidermis aecial primordia are formed. The spermatial nuclei through the receptive hyphae reach the aecial primordia. Dikaryotization occurs (Fig. 26).

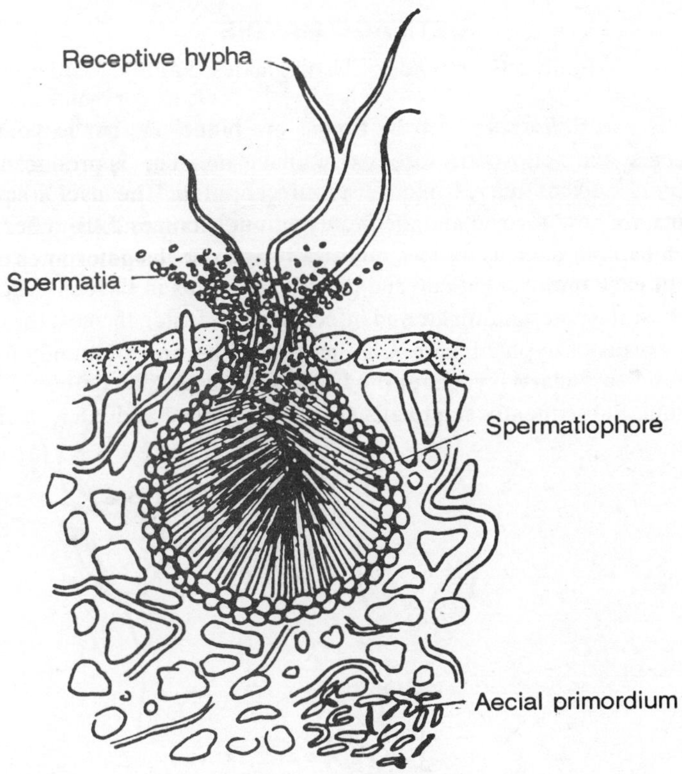

Fig. 26 Spermogonium

Aecia and Aeciospores

Dikaryotic aecial primordia in the barberry leaf grow into aecia and aeciospores. Aeciospores are binucleate spores. The aeciospores are formed in chains and break through lower epidermis (Fig. 27). The aeciospores are carried by wind. They cannot infect barberry and can infect only wheat. When the spores fall on wheat leaves, they germinate, penetrate, multiply and produce binucleate uredospores. The life cycle is repeated.

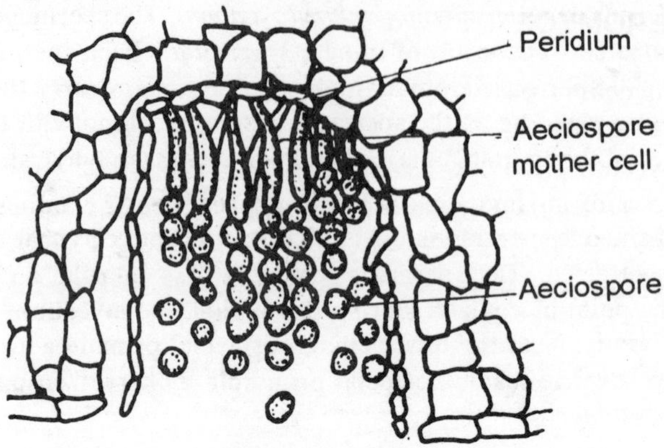

Fig. 27 Aecium

USTILAGO MAYDIS

(Sub class: Teliomycetidae; Order : Ustilaginales; Family: Ustilaginaceae)

It causes maize smut disease. *Teliospores* of the fungus are binucleate in the young stage. Karyogamy (fusion of nuclei) occurs and teliospores become uninucleate. The spores germinate by producing promycelium. The diploid nucleus moves into the promycelium and meiosis takes place. Four haploid nuclei are formed. Septa are now formed and the promycelium becomes four-celled, each cell containing a haploid nucleus. Each haploid nucleus divides mitotically and one daughter nucleus migrates into a bud developing at the side of each promycelial cell and the other remains in the cell. The uninucleate buds are the *basidiospores.* The basidiospores germinate and infect the host. Inside the host the uninucleate mycelium becomes dikaryotic by means of hyphal fusion. Clamp connections are commonly found in the secondary mycelium. The dikaryotic secondary mycelium produces binucleate teliospores. . The teliospores are disseminated by the wind. Subsequently smut balls *(sori)* are formed and later, haploid teliospores are formed (Fig. 28).

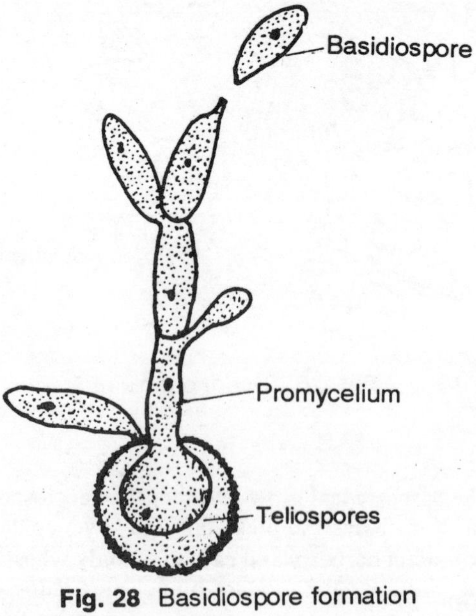

Fig. 28 Basidiospore formation

FORM-CLASS : DEUTEROMYCETES

Some fungi have septate mycelium and reproduce only by producing conidia. Sexual reproduction has not yet been discovered. Whenever the sexual stage of them is discovered, they have been found to belong to Ascomycetes or Basidiomycetes. Till the sexual stage is discovered, the fungi are called 'Fungi Imperfecti' and they are grouped into a form-class (tentative class) called Deuteromycetes under the subdivision Deuteromycotina.

The form-class Deuteromycetes consists of three Form-subclasses. They are : Blastomycetidae, Coelomycetidae and Hyphomycetidae.

Many important pathogens belong to Deuteromycetes. Few examples are given below:

Form subclass Coelomycetidae:

Form order: Sphaeropsidales, Family: Sphaeropsidaceae: (The asexual structure is a pyenidium).

Phoma, Macrophomina, Phomopsis, Phyllosticta, Ascochyta, Diplodia, Septoria

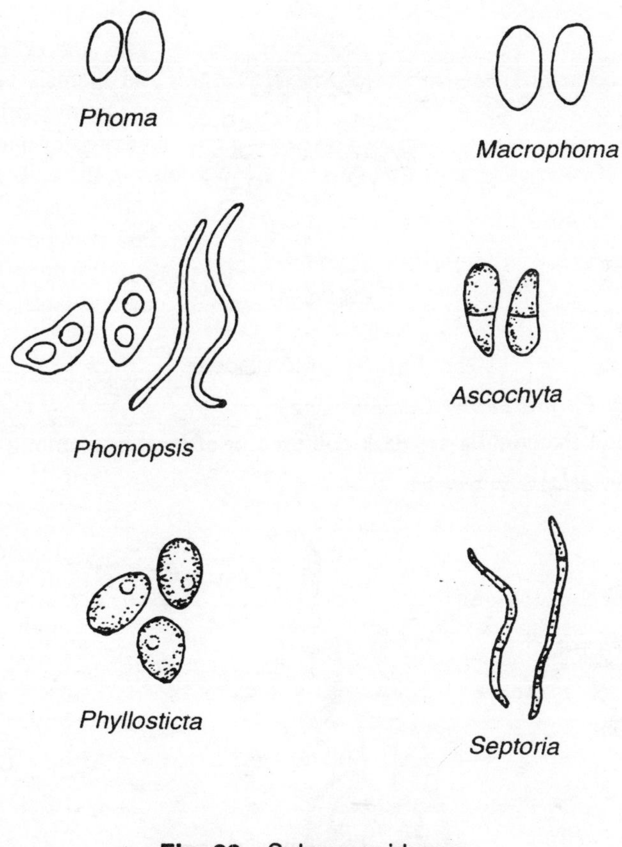

Fig. 29 Sphaeropsidaceae

Form-order: Melanconiales; Form-Family: Melanconiaceae: The asexual structure is acervulus. *Colletotrichum, Gloeosporium and Pestalotia.*

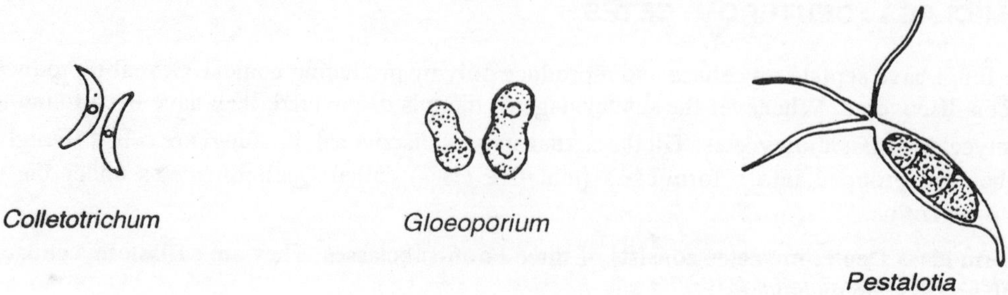

Fig. 30 Melanconiaceae

Form subclass: Hyphomycetidae

Form order: Moniliales; Form-Family: Moniliaceae:

(The conida are hyaline in this family)

Pyricularia, Verticillium, Botrytis

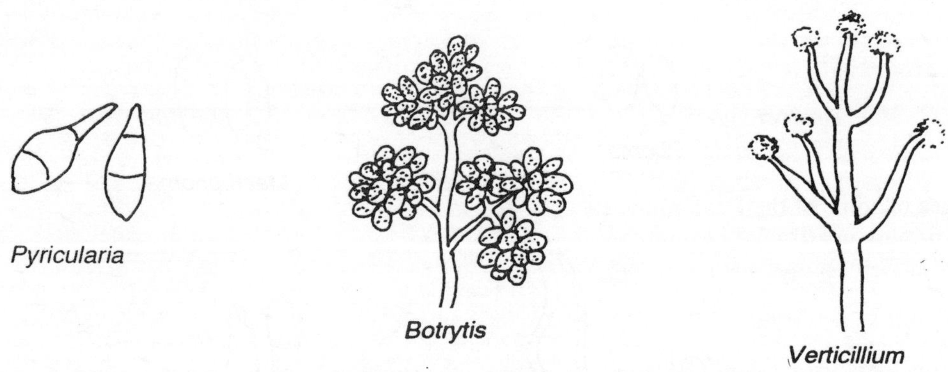

Fig. 31 Moniliaceae

Form order: Moniliales; Form-Family: Dematiaceae

(Both the hyphae and the conidia are dark coloured or at least one among them are dark coloured)

Helminthosporium, Alternaria, Cercospora

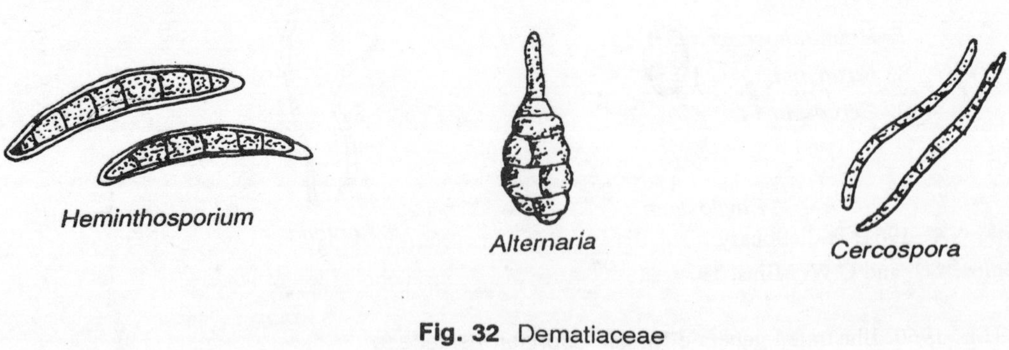

Fig. 32 Dematiaceae

Form order: Moniliales; Form-Family: Tuberculariaceae

(The asexual reproductive structure is a sporodochium; see Fig. 5):

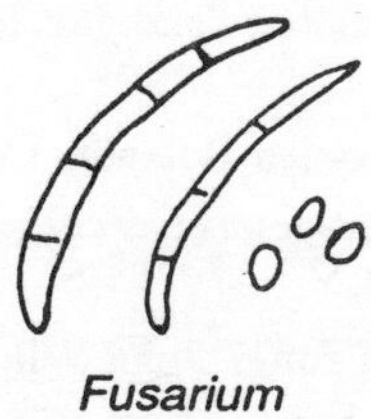

Fig. 33 Tuberculariaceae

Form order: Agonomycetales (Mycelia sterilia)

No sexual and asexual spores are detected. Sclerotia are the only resting spores in the fungi. Examples are *Rhizoctonia* and *Sclerotium.*

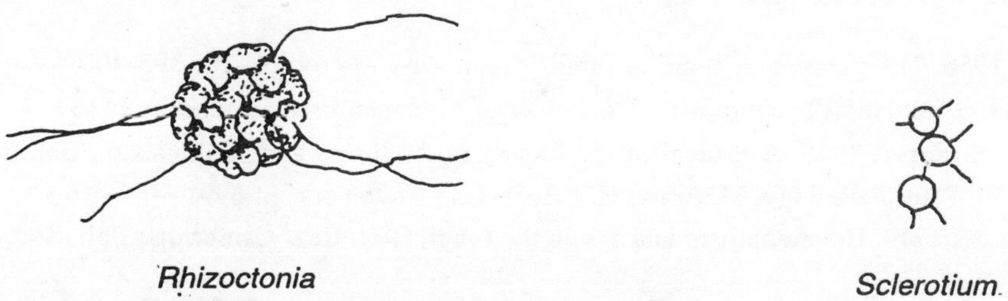

Fig. 34 Agonomycetales

Perfect stages of some of the Deuteromycetes have been detected. But the sexual spores are produced very rarely or only in artificial medium. Hence even those fungi whose sexual stage has been reported, ar placed only in Deuteromycetes because of rarity in the occurrence of sexual stage.

Imperfect stage		*Perfect stage*
1. *Colletotrichum*	:	*Glomerella*
2. *Septoria*	:	*Mycosphaerella*
3. *Oidium*	:	*Erysiphe*
4. *Aspergillus*	:	*Eurotium, Emericella, Sartoya*
5. *Penicillium*	:	*Talaromyces, Eupenicillium*
6. *Pyricularia*	:	*Magnaporthe grisea*
7. *Helminthosporium*	:	*Cochliobolus, Pyrenophora*
8. *Fusarium*	:	*Gibberella, Nectria*
9. *Cercospora*	:	*Mycosphaerella*

REFERENCES

1. Ainsworth, G.C. 1961. A dictionary of the fungi. CMI, Kew, Surrey, p. 547
2. Alexopoulos, C.J. and C.W. Mims, 1979. Introductory Mycology. John Wiley & Sons, Inc., New York, p. 609.
3. Barnett, H.L. 1960. Illustrated genera of imperfect fungi. Burgess publishing Co., Minneapolis, p. 225
4. Bessey, E.A. 1950. Morphology and taxonomy of fungi. The Blakiston Co., Philadelphia, p. 791
5. Butler, E.J. and S.G. Jones, 1949. Plant Pathology. MacMillan and Co., London, p. 979

6. Cantino, E.C. and C.F. Turian, 1959. Physiology and development of lower fungi (Phycomycetes). Annu. Rev. Microbiol, **13**: 97-124
7. Chester, K.S. 1946. The cereal rusts. Chronica Botanica Co., Waltham, Mass, p. 269
8. Chupp, C. 1953. A monograph of the fungus genus *Cercospora.* Published by the author, Ithaca, New York, p. 667
9. Cochrane, V.W. 1958. The physiology of fungi. John Wiley and Sons, New York, p. 524
10. Cook, W.R.I. 1933. A monograph of the Plasmodiophorales. Arch. Protistenk, **80**: 179-254
11. Cummins, C.B. 1959. Illustrated genera of rust fungi. Burgess publishing Co., Minneapolis, p. 131
12. Fischer, G.W. and C.S. Holton. 1957. Biology and control of the smut fungi. The Ronald Press Co., New York, p. 622
13. Gilman, J.C. 1957. Soil fungi. Iowa State University Press, Ames, p. 450
14. Hawker, L.E. 1957. The physiology of reproduction in fungi. Cambridge University Press, Cambridge, p 128
15. Pontecorvo, G. 1956. The parasexual cycle in fungi, Annu. Rev. Microbiol, **10**: 393-400
16. Sparrow, F.K. 1960. Aquatic Phycomycetes. University of Michigen Press, Ambor, p. 1187
17. Thom, C. and K.B. Raper. 1945. A manual of the Aspergilli. Williams and Wilkins Co., Baltimore
18. Walker, J.C. 1950. Plant Pathology. McGraw, Hill Book Co., New York, p. 699
19. Whitehouse, H.L.K. 1949. Heterothallism and sex in the fungi. Biol. Rev. Cambridge Phil. Soc. **24**: 411-447
20. Wolf, F.A. and F.T. Wolf. 1947. The fungi. John Wiley and Sons, New York, p. 438.

3

THE PATHOGENS - BACTERIA

Bacteria belong to prokaryota which encompasses organisms with a primitive type of nucleus lacking a clearly defined membrane while fungi and algae (excluding blue green algae) belong to eukaryota which encompasses organisms possessing a well-defined nuclear membrane and chromosomes. The bacteria are smaller than fungi and measure about 0.5 to 1.0 × 2.0 to 5.0 μ.

SHAPE AND ARRANGEMENT OF BACTERIAL CELLS

Shape of the bacteria varies. The rod shaped bacteria are called bacillus. The spherical or ellipsoidal bacterial cells are called cocci. The spherical cells are arranged in different forms. When pair of cells occurs, it is named as *Diplococcus.* When the cells are in rows, the bacterium is called *Streptococcus.* Bacterium with clusters of cells resembling bunches of grapes is called *Staphylococcus.* The bacterium with cubes of cells is called *Sarcinae.* When the bacterium is spiral shaped, it is called *Spirillum* (Fig. 35).

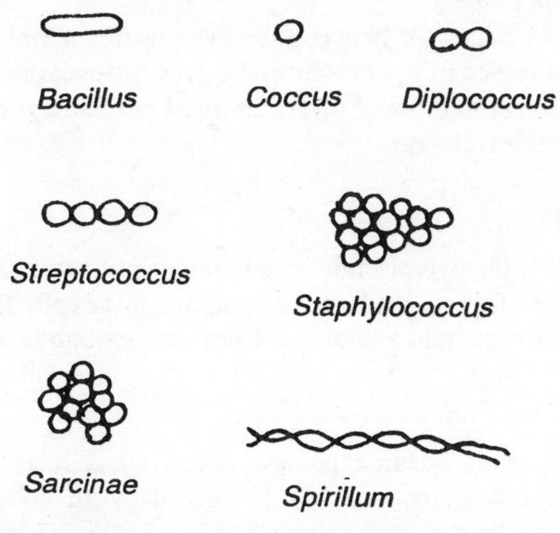

Fig. 35 Shape of bacterial cells

STRUCTURE OF BACTERIA

A bacterial cell consists of a cell wall and a compound membrane enclosing protoplasm. Inside the protoplasm, nucleus, vacuoles, mesosomes, ribosomes, polysaccharides, lipids, metachromatic granules and spores are found. Externally bacterial cells may show capsules, fimbriae, pili and flagella (Fig. 36).

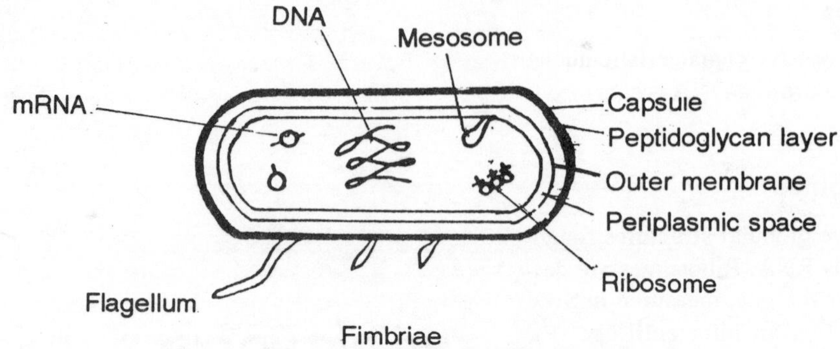

Fig. 36 Structure of a bacterial cell

(i) Slime Layer and Capsule

Some bacteria form an extracellular, slimy or gelatinous, polysaccharide like material as the outermost layer. This material may remain firmly adherent as a discrete covering layer of each cell, or it may part freely from the cell as it is formed. In the former it is called a capsule and in the latter, it is called a free slime. The capsule is a part of the bacterial cell while the slime is a secretion.

(ii) Cell Wall

The bacterial cell contains a characteristic cell wall. Cell wall of higher plant cells is composed of cellulose and that of fungi is largley chitin. The cell wall of bacteria is composed of a peptidoglycan. The peptidoglycan is also called mucopeptide or glycopeptide or murein. Murein is composed of six different components viz. N-acetylglycosamine, N-acetylmuramic acid, L-alanine, D-alanine, D-glutamic acid and either L-lysine or meso-diaminopimelic acid.

The rigid peptidoglycan layer is located between the cytoplasmic membrane and an outer multiple - tract layer. The latter layer is composed of lipoprotein and lipopolysaccharide complexes. It is common in gram-negative bacteria. The external multitrack layer, the rigid peptidoglycan layer and the cytoplasmic membrane constitute the bacterial envelope.

(iii) Cytoplasmic Membrane

Immediately below the cell wall is the cytoplasmic membrane or plasma membrane. It is semipermeable. It controls the passage of nutrients of waste products into and out of the cell. The membrane contains about 75 per cent protein, 20 to 30 per cent lipid and about 2 per cent carbohydrate.

(iv) Mesosomes

Besides cytoplasmic membrane, many bacteria possess other intracellular membrane systems such as mesosomes or chondrioids. The mesosome structure is formed by an invagination of the cytoplasmic membrane. Mesosomes serve for compartmentalization and integration of biochemical systems.

(v) Cytoplasm

The cell material contained within the cytoplasmic membrane can be divided into the cytoplasmic area rich in ribonucleic acid (RNA), chromatinic area (nuclear area) rich in deoxyribonucleic acid (DNA) and the fluid portion with dissolved nutrients.

(vi) Nucleus

Bacteria do not have characteristic nucleus. They contain bodies within the cytoplasm that are regarded as a nuclear structure and DNA is confined to this area. Since, it is not a discrete nucleus, it is called 'chromatin body'.

(vii) Ribosomes

Ribosomes are globular structures found in the cytoplasm. They are composed of about one-third protein and two-thirds RNA. Ribosomes are designated 30s, 50s, 70s, etc. depending upon their size. The size is determined by the rate, measured in Svedberg unit (s) at which a particle sediments when it is centrifuged at high speed in an ultracentrifuge. Ribosomes act in clusters and they are called polyribosomes or polysomes. Polysomes may contain 3 to 70 ribosomes depending upon the bacteria. Ribosomes are the sites of protein synthesis.

(viii) Vacuoles

Vaculoes are cavities in the cytoplasm and contain a fluid called cell sap. As the cell approaches maturity, some of the water-soluble reserve food materials manufactured by the cell get dissolved in the sap. Insoluble constituents precipitate out as cytoplasmic inclusion bodies.

(ix) Cytoplasmic Inclusions

Volutin: They are metachromatic granules. They are found to localize in the vacuoles of mature forms. They contain inorganic polyphosphate, lipoprotein, RNA and magnesium. They may serve as phosphate storage structures.

Glycogen: It accumulates in the cytoplasm at the ends of the cell in the form of granules.

Fat globules: Lipids are found in the cytoplasm of bacteria in the form of fat globules.

(x) Flagella

Some of the bacteria possess flagella which are useful for motility of the bacteria. Some bacterial cells have polar flagella - a single flagellum or a tuft of flagella at one end. The bacteria belonging to Eubacteriales have peritrichous flagellus - flagella at all the sides.

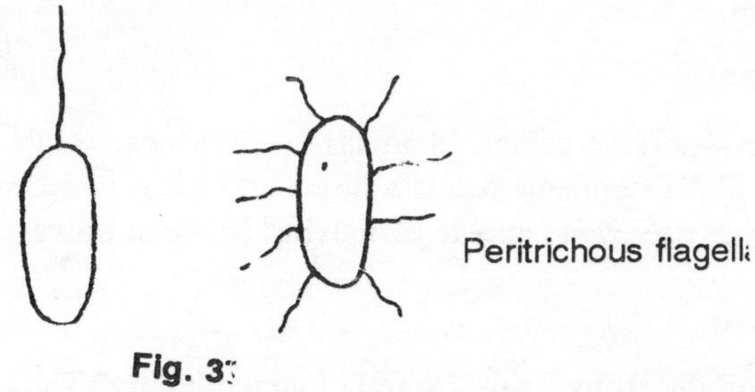

Fig. 3

(xi) Endospores

Some bacteria transform themselves into small ovals or spheres which are highly resistant to adverse conditions. They are called as spores. They are produced intracellularly. Bacterial spores have dipicolinic acid which is absent in vegatative cells. Dipicolinic acid is found only in bacterial spores.

(xii) Fimbriae

Fimbriae are hair-like structures that are observed as surface appendages on some bacteria. Fimbriae are common in plant pathogenic bacteria.

(xiii) Pili

Pili are also hairlike structures found in some bacteria. They serve as adsorption organs for bacteriophages. Pili are considered to be sex organs. They mediate conjugation (mating) of bacteria.

(xiv) Protoplast, Sphaeroplast, L forms

Bacterial cell lacking cell wall is called protoplast. The naked protoplast may synthesize a portion of cell wall material. Such a protoplast is called *sphaeroplast.* L forms are non-rigid bacterial forms. They are soft protoplasmic elements without defined morphology which can be propagated indefinitely on solid medium.

Protoplasts, spheroplasts and L forms are all bacterial forms devoid of the rigid cell wall.

(xv) Chromosomes

Chromosomes consist of nucleoproteins (proteins and nucleic acids). They are the carriers of genes (the factors that control inherited traits). Genes are nucleic acids and more specifically DNA. Genes are linked and aligned on the chromosome. Chromosome maps show the location of various genes. Chromosomes are located in the chromatin body of the bacteria.

(xvi) Plasmid

Some bacteria have plasmids in the cytoplasm. Plasmids are extrachromosomal DNA capable of autonomous replication. They are dispensible circular DNA strands and they are not capable of integration with the chromosome.

(xvii) Episome

Episomes are similar to plasmids. They are also autonomous and dispensible genetic elements. But unlike plasmids, episomes, can exist even integrated with the chromosome. Generally the bacteria containing plasmids do not have episomes and vice versa.

(xviii) Transposon

Transposons are transposable genetic elements — (transpose = alter the positions of;interchange). Transposons are mobile DNA segments that can insert into a few or several sites in a genome (genome = set of genes). Transponsons are even capable of moving between procaryotes and eucaryotes.

(xix) Siderophore

Siderophores are produced by many bacteria extracellularly. They are low molecular weight iron (III) - transport agents. The function of siderophores is to supply iron to the cell. Siderophores selectively complex with iron (III) (ferric iron) with very high affinity. The siderophore isolated from *Pseudomonas* sp. is called *pseudobactin.*

(xx) Bacteriocin

Bacteriocins are non-replicating, bactericidal protein - containing substances which are produced by certain strains of bacteria and are active against some other strains of the same or closely related species. Bacteriocins include complete phage-like particles and small headed "killer particles" containing both protein and nucleic acid as well as low molecular weight substances sensitive to proteases and nucleases.

Bacteriocins produced by different bacteria are named differently. Some bacteriocins are named after the genus of the bacteria. For example, bacteriocins produced by *Agrobacterium* species are named as *Agrocins.* Some bacteriocins are named after the species of the bacteria. For example, bacteriocins produced by *Pseudomonas syringae* are called *syringacins.* In some cases bacteriocins are named after the strain number of the bacteria. Bacteriocin produced by *Agrobacterium radiobacter* strain 84 is called bacteriocin 84.

High-molecular weight, trypsin-resistant bacteriocins are produced by strains of *Corynebacterium michiganense, Erwinia chrysanthemi, Pseudomonas phaseolicola, P. syringae, P. glycinea* and *P. solanacerum.* Low molecular weight bacteriocins are produced by some strains of *P. syringae, P. phaseolicola, P. glycinea* and *C. michiganense.* Particulate contractile bacteriocins are produced by *P. syringae* 4-A and *P. syringae* W-1. Killer particles are produced by strains 396 and 0362 of *Agrobacterium tumefaciens.*

Bacteriocins are useful in the control of diseases. Bacteriocins produced by avirulent isolates are exploited to control diseases caused by virulent isolates of the same bacterium. The use of bacteriocins in control of diseases is discussed in Chapter 18.

REPRODUCTION OF BACTERIA

(i) Fission

The predominant mode of reproduction in bacteria is binary fission. The bacterial cells divide into two daughter cells. It is asexual reproduction.

(ii) Conjugation

It is a sexual reproduction. Genetic material of one cell is transferred to another cell during conjugation. The two cells are genetically different. The donor cell transfers part of its genome (set of genes) to the recipient cell. Donor strains are designated F^+ or Hfr (males). The recipient strains are designated F^- (females).

The F factor is called sex factor. It is a type of episome. Episomes are genetic elements which can exist in two alternative states; autonomous in the cytoplasm, or incorporate in the chromosome. In the former condition, the episome is free to multiply independently of cell division; in the latter, it is replicated only when the chromosome is replicated. Cytoplasmic episomes occur in F^+ cells while the episome is incorporated into the chromosomes of Hfr (high frequency) cells. F factor controls the ability of the cells to act as gene donors. Hfr strains exhibit high frequency of recombination for characters other than F^+.

In the process of conjugation the donor cell (Hfr or F^+) injects a chromosomal thread into the recipient cells, and the genes along this thread enter in a definite order, one after another, as they occur in a particular donor strain.

Genetic recombination refers to any process leading to the formation of a new individual which derives some of its gene from one parent and some from another, genetically different, parent. The recipient cell which recieves the genes from the donor is called a recombinant. Conjugation is one type of genetic recombination.

(iii) Transformation

DNA from one type of bacterium is incorporated into the genetic make-up of another organism during transformation. A new individual which derives some of its genes from two genetically different parents is formed during this process.

(iv) Transduction

The virus which infects a bacterium is called '*bacteriophage*'. Bacteriophage grows within a bacterial cell. The infected bacterial cell bursts and the bacteriophage particles are liberated. Some of the bacteriophage particles released carry some genetic material of the host bacterial cell. They enter into another bacterial cell and the genetic material brought into the cell by the virus is incorporated into that of the bacteria. By this process a new genotype of the bacterium arises and this process is called transduction.

GRAM STAINING

The bacteria are grouped as gram-positive and gram-negative based on gram stain reaction. Most of the plant pathogenic bacteria are gram-negative. Hence gram staining is useful to distinguish plant pathogens.

To do gram staining the bacterial smear is prepared and subjected to the following solutions in the order listed: crystal violet, iodine solution, alcohol, and saffranin. Gram-positive bacteria retain the crystal violet and hence appear deep violet in colour. Gram-negative bacteria lose the crystal violet and counterstained by the saffranin, and hence appear red in colour.

Gram negative bacterial cell walls contain more of lipids (about 20% of cell wall material is lipid). Alcohol treatment extracts the lipid and the crystal violet-iodine complex is also extracted. The crystal violet-iodine complex is retained in gram positive bacteria during alcohol treatment.

Bacteriophage

Viruses that infect bacteria are called bacteriophages. Bacteriophages are shaped like tadpoles with round, oval or polyhedral heads and a hollow tail with specialized fibrils at their tip (Fig. 38).

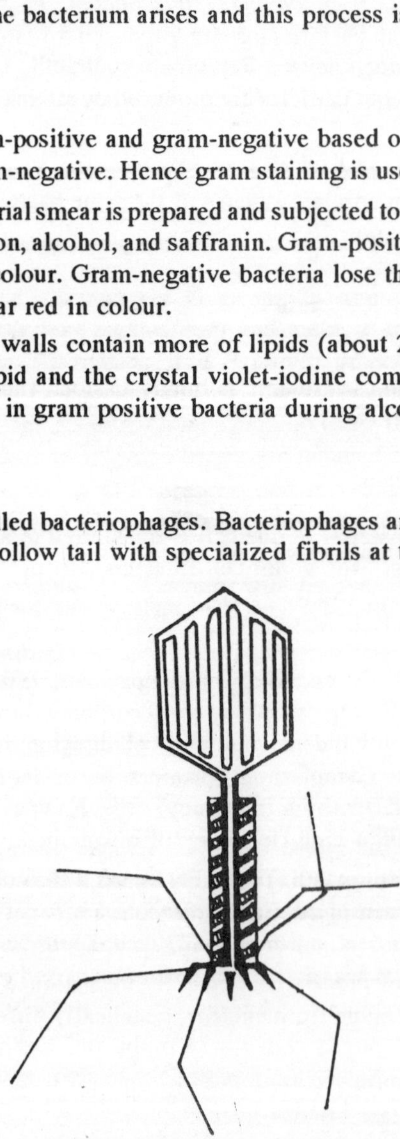

Fig. 38 Bacteriophage

The head is composed of DNA surrounded by a protein membrane which covers the tail. The protein coat gives protection to the bacteriophage in its extracellular environment. The tail fibres attach to the cell wall of the bacteria. They secrete an enzyme which penetrates the cell membrane to permit entry of the virus DNA. The DNA enters through a pore and induces the bacterial cell to manufacture more virus particles. When bacteriophage particles multiply quickly, the bacterial cell dies. At the death of the cell, the bacteriophages are released. The bacteriophages infect new bacterial cells. Bacteriophages survive only in bacterial cells and in the absence of them they perish.

CLASSIFICATION OF PLANT PATHOGENIC BACTERIA

The bacteria belong to the class Schizomycetes. Plant pathogenic bacteria belong to 3 different orders 5 families, and 6 genera.

(i) Order: Pseudomonadales; Family: Pseudomonadaceae

This family is characterized by bacterial cells with single polar flagella; the bacteria are gram negative.

Genera:

Pseudomonas: It does not produce non-water soluble pigment; produces soluble pigment which is not yellow. No acid is produced from lactose.

Xanthomonas: Cultures develop an yellow non-water soluble pigment; produce acid from lactose.

(ii) Order: Eubacteriales; Family: Rhizobiaceae

The genus *Agrobacterium* is pathogenic on plants. The cells are rod shaped, gram negative and sparsely flagellated. The colonies are white.

(iii) Order: Eubacteriales; Family: Enterobacteriaceae

The genus *Erwinia* belongs to this family. The bacterial cells are gram negative and motile with peritrichous flagellum. They produce acid from sugars.

(iv) Order: Eubacteriales; Family: Corynebacteriaceae

The genus *Corynebacterium* is pathogenic on plants. The bacterial cells are gram positive. *Corynebacterium* is the only gram positive plant pathogen while others are gram negative. The bacterial cells are pleomorphic (L and Y forms) rod that show the charactristic arrangement produced by snapping division. They are motile rods.

(v) Order: Actinomycetales; Family: Streptomycetaceae

The genus *Streptomyces* has plant pathogenic species (*Streptomyces scabies* - potato wart disease). It produces conidia in aerial hyphae in chains. It has characteristic branching mycelium and spores are formed by fragmentation of the plasma within straight or spiral shaped, spore-bearing hyphae *(oidiospores)*.

Important Plant Pathogens

(i) **Pseudomonas**

Pseudomonas solanacearum: Tomato, Potato, Banana, Brinjal wilts.

P. syringae pv. *phaseolicola:* Halo blight of beans

P. tabaci: wild fire of tobacco

(ii) Xanthomonas

Xanthomonas campestris: black rot of cabbage
X. campestris pv. *oryzae:* bacterial blight of rice
X. campestris pv. *malvacearum:* black arm of cotton
X. campestris pv. *citri:* citrus canker

(iii) Corynebacterium

Corynebacterium michiganense: tomato bacterial canker
C. sepedonicum: leaf spot and wilt in many crops
C. insidiosum: leaf spot in many crops
C. rathayi: gummosis in many crops
C. tritici: wheat tundu disease

(iv) Erwinia

Erwinia carotovora: soft rot of many vegetables.
E. amylovora: fire blight of apple

(v) Agrobacterium

Agrobacterium tumefaciens: crown gall in many crops

(vi) Streptomyces

Streptomyces scabies: potato scab.

MODE OF ENTRY

Bacteria cannot penetrate host tissues as they don't have germtubes and appressorium. They enter the host through different types of natural opening.

(i) Hydathodes : They are gland cells at leaf edges. *X. campestris* enters through hydathodes
(ii) Floral parts : *Erwinia amylovora* enters through nectaries of apple flowers.
(iii) Lenticels : *Streptomyces scabies* enters through lenticels of potato tuber.
(iv) Stomata : *Pseudomonas tabaci* and *Xanthomonas malvacearum* enter through stomata
(v) Wounds : *X. campestris* pv. *oryzae* enters into rice leaves through wounds caused by wind.

MODE OF SPREAD

(i) **Seed:** Many bacteria are transmitted through seed. Some examples are: black arm of cotton caused by *X. campestris* pv. *malvacearum* and rice bacterial blight caused by *X. campestris* pv. *oryzae*.

(ii) **Soil:** The bacteria survive in soil for short time only. However *Pseudomonas solanacearum* is a soil inhabitant and causes wilt diseases in many crops like potato, tomato and banana.

(iii) **Insect:** Honey bees spread *E. amylovora* from blossom to blossom in apple. Flies and ants feed on the sticky exudates from canker lesions and spread the bacteria from the stem to the blossom of apple. The corn flea beetle, *Chaetocnema pulicaria* carries *Xanthomonas stewartii* , the corn wilt pathogen inside its body. The bacteria supply digestive enzymes to the insect and hence the bacteria can be called symbionts.

(iv) **Nematode:** The nematode, *Anguina tritici* carries the bacterium, *Corynebacterium tritici* into the impermeable tissues of wheat kernels. *C. tritici* causes tundu disease of wheat. The nematode, *Meloidogyne incognita* var. *acrita* helps *Pseudomonas solanacearum* to cause wilt of tomato.

(v) **Crop debris:** Infected leaves and debris in soil harbour the bacteria for long time. Example: *X.campestris* pv. *malvacearum,* the black arm of cotton pathogen.

(vi) **Collateral hosts:** The bacterial pathogens have wide host range and many weeds are affected. The weeds serve as collateral hosts.

(vii) **Self-sown and volunteer crops:** They act as carrier of the bacteria from season to season. Example - Bacterial blight of rice.

DISEASE SYMPTOMS

(i) **Leaf spots:** Water soaked, circular or irregular or angular spots are common.
Example — Black arm of cotton (*X. campestris* pv. *malvacerum*)

(ii) **Leaf blight:** The entire leaf may be blighted
Example — Paddy bacterial blight (*X. campestris* pv. *oryzae*)

(iii) **Cankers:** Corky outgrowths are formed on leaves, twigs and fruits.
Example — Citrus canker (X. *campestris* pv. *citri*)

(iv) **Wilt:** The bacteria enter vascular tissues and cause wilt.
Example — Tomato wilt *(Pseudomonas solancearum)*

(v) **Soft rot:** The bacteria enter through wounds, produce pectolytic enzymes and disintegrate the host tissues.
Example — Soft rot of vegetables *(Erwinia carotovora)*

(vi) **Tumours and galls:** The bacteria enter through wounds and cause hyperplasia and hypertrophy of host cells
Example — Crown gall of fruit trees and vegetables *(Agrobacterium tumefaciens)*

(vii) **Scab:** The surface of affected tissues becomes rough and the affected surface is raised above due to abnormal proliferation of the cells in the epidermis. The surface has rusty appearance and deeply pitted and corky wounds.
Example — Potato scab *(Streptomyces scabies)*

PATHOGENICITY FACTORS

The important events in pathogenesis of bacteria are recognition, multiplication and translocation of the pathogen and ultimately, stimulation or inhibition of the host. The known determinants of bacterial pathogenicity are cell envelope components (capsule and extracellular polysaccharide (EPS)), flagella, pili, fimbriae, extracellular enzymes, toxins and indole acetic acid (IAA).

(I) Capsule

Capsule is the outermost component of the cell periphery. In *Erwinia amylovora,* non-capsular strains obtained as laboratory mutants or as naturally occurring strains are generally non-pathogenic. Non-capsular mutants of *Pseudomonas solanacearum* are avirulent. Capsule secretes extracellular polysaccharide (EPS). The amount of EPS production has been correlated with virulence of bacterial pathogens such as *Xanthomonas campestris* pv. *oryzae, X. campestris* pv. *campestris, P. solanacearum,* and *Erwinia stewartii.*

(II) Flagella

Flagella help the bacteria for movement. Motile *Pseudomonas syringae* pv. *phaseolicola* causes more infection loci than paralyzed mutants.

(iii) Fimbriae

Fimbriae are commonly found in plant pathogenic bacteria such as *P. solanacearum, P. syringae* pv. *syringae, Agrobacterium tumefaciens* and *Erwinia carotovora.* They function in adherence to surfaces of plant cells.

(iv) Enzymes

Bacterial pathogens produce pectolytic enzymes which facilitate tissue disintegration. Example: *Erwinia carotovora* causing soft rot of vegetables.

(v) Toxins

Pseudomonas syringae pv. *phaseolicola* and *P. syringae* pv. *tabaci* are known to produce toxins. The mode of action of toxins in disease development is discussed in Chapter 11.

(vi) Indole acetic acid

IAA is produced by many plant pathogenic bacteria. It modifies the host metabolism and induces hypertrophy. Example: *Agrobacterium tumefaciens* produces IAA in infected tissues.

DISEASE MANAGEMENT

The principles of bacterial disease management are similar to fungal disease management as discussed in Chapters 15 to 20. Only some specific methods pertaining to bacterial diseases are given below:

1. **Exclusion:** Many bacterial pathogens are seed-borne. It is very difficult to eradicate them from infected seed. Hence seeds from infected areas should not be imported into the new areas. Seed health testing should be strictly carried out. Seeds from disease-free fields alone should be used.
2. **Eradication:** Bacterial diseases are difficult to be controlled. Hence, it is better to eradicate the disease from a locality itself. In U.S.A. citrus canker has been eradicated by cutting down all the citrus trees in the country.
3. **Cultural practices:** Roguing of infected plants, weeding the possible alternate hosts and removal of volunteer plants should be done to reduce the inoculum.
4. **Resistant varieties:** Growing resistant varieties is the best way of protection against bacterial pathogens. Several resistant varieties against bacterial blight of rice and black arm of cotton have been evolved in India. However many races of the pathogen exist in nature and breakdown of resistance in the cultivars is common. Hence new resistant varieties should be introduced every now and then.
5. **Biological control:** Bacterial diseases can be controlled by inoculating avirulent pathogens prior to the inoculation of the pathogen. Hence, it is not useful under field conditions. However bacteriocins isolated from incompatible strains can be sprayed to control bacterial diseases. Syringacin from *Pseudomonas syringae* controls halo blight of bean caused by *P. syringae* pv. *phaseolicola.*
6. **Use of bacteriophages:** Bacteriophages, when applied prior to the bacterial pathogen, effectively control the diseases. However, if they are applied after the infection, the disease is not controlled. Although bacteriophages are potential tools to control bacterial pathogens, they should not be used due to the following reasons:
   (i) Bacteriophages may transduce various characters from one bacterium to another. By this way non-pathogen or avirulent strain may become highly pathogenic and virulent.

(ii) Transducing phages can introduce active bacterial genes into the plant cells making the plants sick even in the absence of bacteria.

7. **Antibiotics:** Since bacteria do not have germtube and appressorium, protectant chemicals will be of no use. Antibiotics which are produced by microorganisms will be useful in the control of bacterial diseases. Streptomycin and tetracyclines are commonly used. However large scale use of antibiotics for plant disease control may result in development of streptomycin resistant human and animal pathogens. This will be highly dangerous to the human beings.

REFERENCES

1. Bisset, K.A. 1962. Bacteria, E and S Livingstone Ltd., Edinburgh and London
2. Breed, R.S., E.G.D. Murray and N.R. Smith. 1957. Bergey's Manual of Determinative Bacteriology, 7th Ed. The Williams and Wilkins Co., Baltimore
3. Chatterjee, A.K. and A.K. Vidaver. 1986. Advances in Plant Pathology. Vol. 4. Genetics of Pathogenicity factors: Application to phytopathogenic bacteria. Academic Press, Lond, p., 224
4. Gunsalus, I.C. and R.Y. Stanier. 1960. The Bacteria. Vol. I. Academic Press, Inc., New York
5. Hayes, W. and R.C. Clowes. 1960. Microbial genetics. Cambridge University Press, New York
6. Pelczar, M.J., Jr., and R.D. Raid 1965. Microbiology. McGraw-Hill Book Co., New York, p. 662
7. Perkins, H.R. 1963. Chemical structure and biosynthesis of bacterial cell walls. Bacteriol Rev., 27: 18-25
8. Rangaswami, G. and Soumini Rajagopalan. 1973. Bacterial Plant Pathology. T.N.A.U., Coimbatore, p. 291
9. Salle, A.J. 1974. Fundamental Principles of Bacteriology. Tata McGraw-Hill Publishing Co., Ltd., New Delhi, p. 1094.
10. Verma, J.P. and R.P. Singh. 1983. Control of bacterial diseases through resistance breeding. In: Akhtar Husain, Kishan Singh, B.P. Singh and V.P. Angihotri (ed.) Recent Advances in Plant Pathology, Print House (India) Lucknow, p. 20-47
11. Vidaver, A.K. 1976. Prospects for control of phytopathogenic bacteria by bacteriophages and bacteriocins. Annu. Rev. Phytopathol, 14: 451-465.

4

THE PATHOGENS — VIRUSES

In 1885, Mayer from Holland published his studies on tobacco mosaic and concluded that the disease is sap transmissible. He used the word 'mosaic' for the first time. In 1890, Ivanowaski, a Russian scientist, observed that the causal agent is small enough to pass through a bacterial filter. In 1897, Beijerinck called the contagious agent in the plant sap as 'contagium vivum fluidum' and subsequently in 1898, he used the word 'virus' for the causal agent. Since then several virus diseases have been reported almost in all the field crops.

VIRUSES

Viruses are infectious agents, not visible under the light microscope (submicroscopic) and small enough to pass through a bacterial filter. Viruses do not have a metabolism of their own and depend upon living host cells for multiplication. They have only one type of nuclei acid, either ribonucleic acid (RNA) or deoxyribonucleic acid (DNA). Nucleic acids may be either single stranded or double stranded. The nucleic acids may be single segmented or with a few segments. The nucleic acids are enclosed in a coat of protein. Number of coat proteins may range from one to seven. The viruses may have an extra coat (envelope) and some other constituents. The following are the examples for different types of viruses.

1. RNA viruses

   A. Single stranded

      (i) With one segment

         (a) With one protein in the coat - Tobacco mosaic virus

         (b) With more number of proteins in the coat - Lettuce necrotic yellows virus (5 proteins)

      (ii) With two segments

         (a) With one protein in coat - Tobacco ring spot virus

         (b) With more number of proteins in the coat - Cowpea mosaic virus (2 proteins)

      (iii) With three segments

         (a) With one protein in the coat - Cucumber mosaic virus

         (b) With more number of proteins in the coat - Not known

(iv) With more number of segments

(a) Without envelope	-	Potato virus x (two to four segments with one protein in the coat)
(b) With envelope	-	Tomato spotted wilt virus (four segments with five proteins in the coat)
B. Double stranded	-	Wound tumour virus (with 12 segments and 7 proteins in the coat); Fiji disease virus (with 10 segments and 7 proteins in the coat)

2. DNA viruses

A. Single stranded	-	Maize streak
B. Double stranded	-	Cauliflower mosaic virus

VIRION

The individual, complete and infectious nucleoprotein particles of a virus are called *virions.* In short, virus particles are called virions. They are also called *nucleocapsids.* The protein subunits form the protein coat, shell or capsid and they are called structural units. The clusters of such units are called morphological units or *capsomers.* The envelope is called as *peplos.*

SATELLITE VIRUS

Satellite virus has a coat of its own but a small RNA genome. It needs another virus to supply replicase and probably other enzymes. It acts as a parasite of plant-parasitic virus. One example is tobacco necrosis satellite virus.

SATELLITE RNA

Satellite RNA has no coat protein. It is associated with particles of some viruses. It completely depends upon those viruses for multiplication. The viruses which help multiplication of satellite RNA are called helper viruses. Satellite RNA aggravates or reduces the symptom development induced by the helper virus. This type of association has been reported in cucumber mosaic virus and tobacco ring spot virus.

SATELLITE RNA DEPENDING UPON SATELLITE VIRUS

Satellite tobacco necrosis virus has been shown to contain a small satellite RNA that is dependent on tobacco necrosis virus for replication and on the satellite virus for encapsidation.

VIROID

Viroids are naked nucleic acids without coat protein. They consist of only RNA. These 'miniviruses' are the smallest known agents of infectious disease. They are subviral in size. Some of the viroids are: Citrus excocortis viroid, potato spindle tuber viroid, chrysanthemum stunt and chlorotic mottle viroid, cucumber pale fruit viroid and hop stunt viroid.

VIRUSOID

Some viruses contain a viroid-like satellite RNA in addition to a linear single-stranded molecule of genomic RNA. They are called virusoids. They show little sequence homology with the viroids, but they do show significant homology with a linear satellite RNA associated with tobacco ring spot virus.

MULTIPARTITE VIRUS

Some viruses have different particle types which differ in size or nucleic acids. In some mutipartite viruses, light and heavy particles exist. Only heavy particles are infectious; e.g. Turnip yellow mosaic virus. But in general, all the types of virus particles are simultaneously required for successful infection; e.g. Tobacco rattle virus.

SHAPE OF VIRUS PARTICLES

Virus particles are of different shapes and sizes. They are elongate (tobacco mosaic virus) or isometric (cowpea mosaic virus) or bacilliform (alfalfa mosaic virus; rice tungro bacillus virus) or spherical (rice tungro spherical virus).

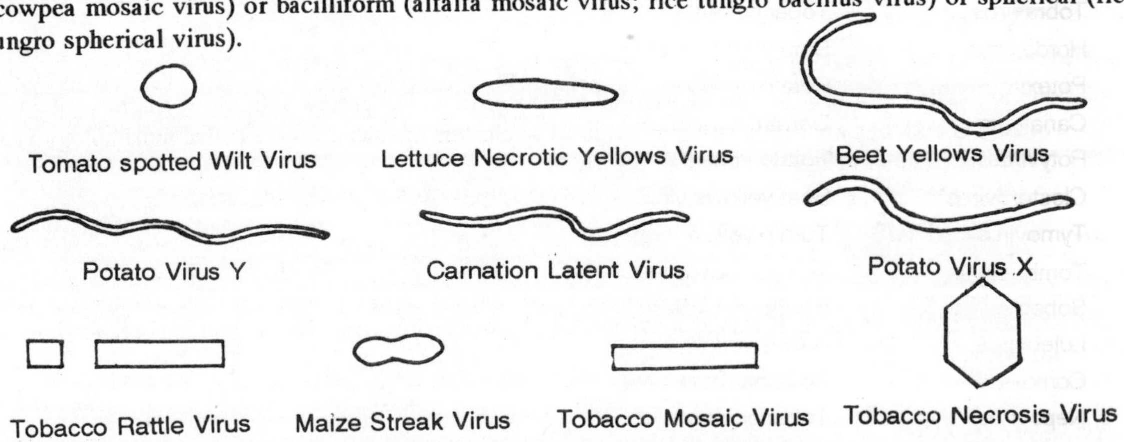

Fig. 39 Different shapes of virus particles

CLASSIFICATION OF PLANT VIRUSES

International Committe on Taxonomy of viruses has classified viruses into different groups. The classification is based on particle morphology and size, naked or enveloped nucleocapsids, number of virion types (multipartite viruses), number of genome fragments, type of nucleic acid (RNA or DNA), and strandedness of nucleic acid (single or double stranded). The group name is mostly based on the type virus included in the group. For example, the group containing the type virus tobacco mosaic virus is called Tobamovirus. Two plant virus groups have been given 'family'status and genera of viruses have also been recognized. The plant virus groups have been summarized and given in Tables 1 and 2.

MODE OF VIRUS INFECTION

Tobacco Mosaic Virus (TMV) infects tobacco and the infection process of the virus has been studied in detail. Once the virus enters into the host, the virus becomes 'naked' by shedding its coat. There is a time lapse between entry and uncoating process. When naked nucleic acid is inoculated, the virus multiplies rapidly compared to the inoculation of intact virus. The place and mechanism of uncoating is not yet fully known.

TMV has one chain of single stranded RNA. This is a + strand which means that some of its information can be directly translated at ribosomes into a virus-specific RNA polymerase, an enzyme required for replication of viral RNA.

Table 1. List of group names of plant viruses approved by International Committe on Taxonomy of Viruses

Group name	Type member	Particles		Nucleic acid		
		Shape	Number differing in size or density	Type	Number of segments	No. of Proteins in capsi protein
Tobamo virus	Tobacco mosaic virus	E	1	sR	1	1
Tobravirus	Tobacco rattle virus	E	2	sR	2	1
Hordeivirus	Barely stripe mosaic virus	E	2-4	sR	2-4	1
Potexvirus	Potato virus X	E	1	sR	1	1
Carlavirus	Carnation latent virus	E	1	sR	1	1
Potyvirus	Potato virus Y	E	1	sR	1	1
Closterovirus	Beet yellows virus	E	1	sR	1	1
Tymovirus	Turnip yellow mosaic virus	I	1+e	sR	1	1
Tombusvirus	Tomato bushy stunt virus	I	1	sR	1	1
Sobemovirus	Southern bean mosaic virus	I	1	sR	1	1
Luteovirus	Barley yellow dwarf virus	I	1	sR	1	1
Comovirus	Cowpea mosaic virus	I	2^c+e	sR	2	2
Nepovirus	Tobacco ring spot virus	I	2^c+e	sR	2	1
Dianthovirus	Carnation ring spot virus	I	2	sR	2	1
Cucumovirus	Cucumber mosaic virus	I	3^c	sR	3	1
Bromovirus	Brome mosaic virus	I	3^c	sR	3	1
Ilarvirus	Tobacco streak virus	I	3	sR	3	1
Geminivirus	Maize streak virus	I	1	sR	1	1
Caulimovirus	Cauliflower mosaic virus	I	1	dD	1	1
Rhabdovirus (Rhabdoviridae family)	Lettuce necrotic mosaic virus	B+En	1	sR	1	5
Reovirus (Reoviridae family)						
Phytoreovirus (Genus)	Wound tumor virus	I	1	dR	12	7
Fijivirus (Genus)	Fiji disease virus	I	1	dR	10	7

E - elongate, I - isometric, B - bacilliform En - enveloped, C - components that differ in density only, e - empty particle, s - single standed, d - double stranded, R - RNA, D - DNA.

Table 2. List of group names of plant viruses to be approved by International Committee on Taxonomy of Virus

Virus name	Particles		Nucleic acid		
	Shape	Number differing in size or density	Type	Number of segments	Number of proteins in capsid protein
Maize chlorotic dwarf virus	I	1	sR	1	1
Tobacco necrosis virus	I	1	sR	1	1
Pea enation mosaic virus	I	2[c]	sR	2	2
Alfalfa mosaic virus	B	4-5	sR	3	1
Tomato spotted wilt virus	I+En	1	sR	4	5

I - isometric, B - bacilliform, En - enveloped, C - components that differ in density only, s - single stranded, R - RNA.

The enzyme attaches to the parental + strand and synthesizes the progeny strand. (-) strand alone is not infectious. The progeny (-) strands act as template for further synthesis of (+) strands. The synthesized (+) strand or a segment of this strand operates as messenger RNA (mRNA) for the production of coat protein. The whole (+) strands are incorporated into the coat proteins and complete virions are synthesized (Fig. 40).

This infection process may slightly vary from virus to virus. For viruses with (-) strand RNA (Rhabdoviruses) special mRNA has first to be transcribed. In case of double stranded RNA viruses like reoviruses, each of the 10 to 12 genome segments may produce its own mRNA. DNA viruses use host polymerases for replication. All viruses have to code for their specific replicase. Coat-protein production is a function encoded in the viral nucleic acid. The coat-protein merely acts as a protective coat during the resting stage of the virus. It also plays a role in recognition during early establishment of infection. Assembly of nucleic acid and coat-protein into virions is almost a spontaneous process when viral nucleic acid and specific proteins meet. The mature virions are inert forms of the virus and act at 'virus spores'. These 'virus spores' help the virus to survive during unfavourable conditions.

The viruses have only limited number of genes. Viroids have only one gene. Tobacco necrosis satellite virus has two genes. The viruses possess about ten genes. The viruses have no information to construct an energetic and protein-synthesizing system of their own. They completely depend upon a living host. Normal building materials and normal energetic system of the plant cells are utilized by viruses. Viruses are small packages of alien genetic information whose message can be read only by susceptible hosts. Virus replication in the host may be very explosive. Even up to 11 per cent of the dry matter of infected tobacco leaves consists of tobacco mosaic virus particles. Virus particles occur throughout their host cells. They are found in the cytoplasm, special organelles and even in the vacuoles. They are found in crystalline array in the nucleoplasm. Bacilliform enveloped viruses accumulate either in vesicular structures of the endoplasmic reticulum or in cavities between the two nuclear membranes. Gemini viruses only occur in the nucleus. Normal metabolism of the host is diverted towards the synthesis of virus nucleoproteins. Hence virus infection normally leads to deficiency symptoms. Severity of disease symptoms mostly depends on the amount of virus produced in the plant.

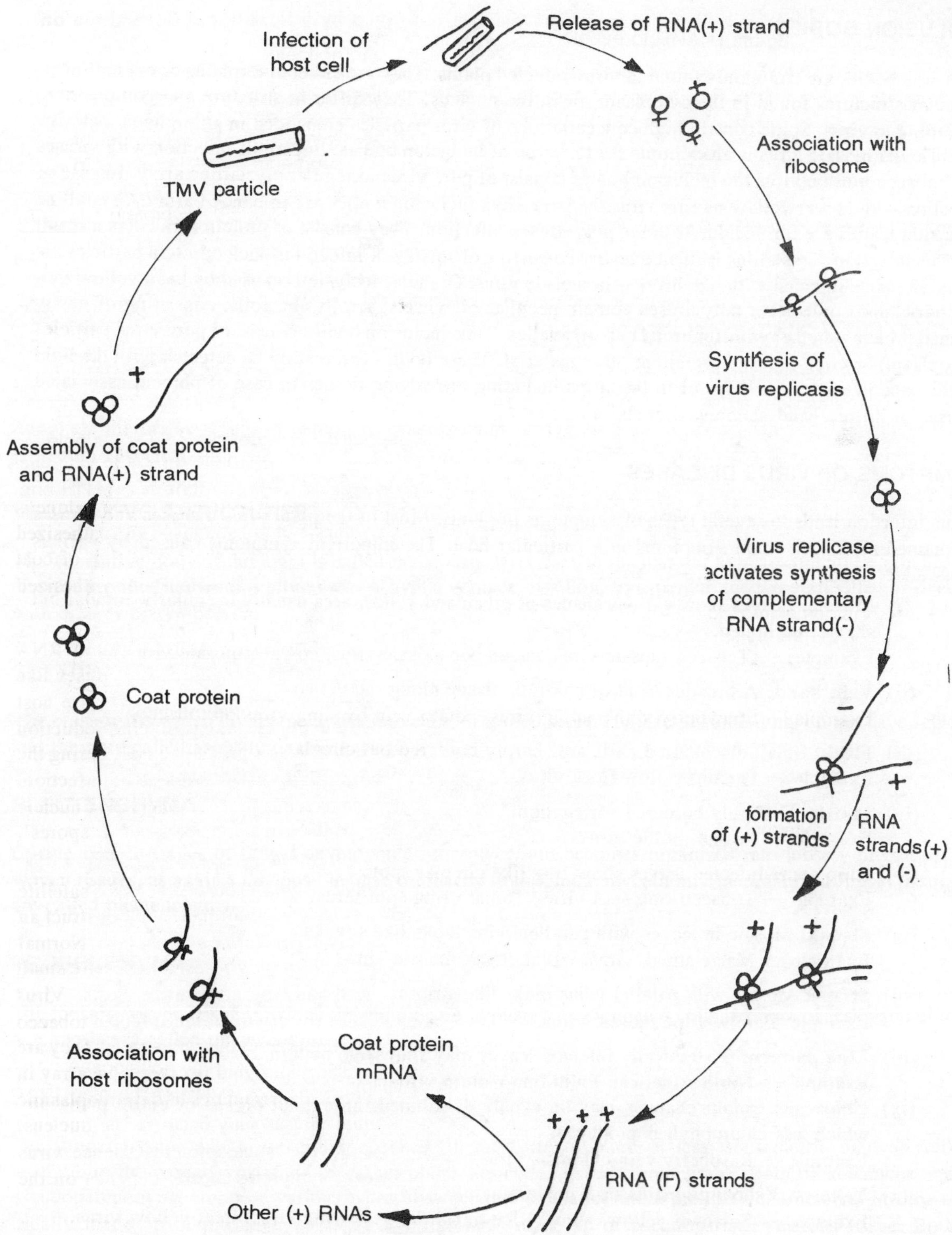

Fig. 40 Infection process of tobacco mosaic virus

INCLUSION BODIES

Inclusion bodies are commonly found in virus-infected plants. They are either amorphous or crystalline or granular structures found in the cytoplasm or in the nucleus. They differ in structure and composition, according to virus. Some contain high concentrations of virus particles embedded in amorphous material. Cauliflower mosaic virus is an example for this type of inclusion bodies. In case of infection with viruses like tobacco mosaic virus the inclusion bodies consist of pure virus in regular crystalline array. In case of infection with clover yellow mosaic virus the pure virus inclusion bodies are in banded array. Crystalline inclusion bodies are formed due to some potyviruses infection. They consist of protein formed as a result of virus infection. Crystalline inclusion bodies consisting of small non-infectious nucleoprotein particles are found in plants infected with redclover vein mosaic virus. Granular inclusions caused by bean yellow mosaic virus and some other potyviruses contain peculiar pinwheels, scrolls and rolls, consisting of newly formed protein together with abnormal cell organelles. Thus inclusion bodies consist of pure virus particles (viroplasm) or virus-induced protein or other material. These bodies can readily be detected with the light microscope in epidermal strips or in the strips including underlying tissues in case of phloem-associated viruses or in free-hand sections after staining.

SYMPTOMS OF VIRUS DISEASES

Virus infection leads to several types of symptoms in plants. Most of the plant virus diseases and viruses are named after their main symptoms in a particular host. The important symptoms caused by various viruses are:

(i) **Mosaic:** Leaves show various shades of green and yellow area usually irregularly angular but sharply delimited.
Examples — Tobacco mosaic virus, cucumber mosaic virus, cowpea mosaic virus.

(ii) **Vein band:** A broader band of chlorotic tissue along the veins.
Example — Strawberry veinbanding virus, potato vein banding virus on tobacco.

(iii) **Fleck:** Small discoloured parts are sharply bordered but circular.
Example — Parsnip yellow fleck virus.

(iv) **Mottle:** Diffusely bordered variegation.
Example — Peanut mottle virus.

(v) **Ring spot:** Infected leaves show ring like circular spots.
Example — Tobacco ring spot virus, Tomato ring spot virus.

(vi) **Streak:** Mosaic in leaves with parallel veins looks like streaks.
Example — Maize streak virus, wheat streak mosaic virus.

(vii) **Stripe:** Mosaic with parallel veins looks like stripes.
Example: Barely stripe mosaic virus.

(viii) **Line pattern:** Systemically infected leaves may show line pattern.
Example — North American Plum line pattern virus.

(ix) **Chlorosis:** Colour changes may be evenly distributed throughout organs or entire plants in which less chlorophyll is produced.
Example — Raspberry vein chlorosis virus.

(x) **Yellows:** Yellow pigments may predominate.
Example — Solanum yellows virus, Rice transitory yellowing virus, Beet yellow virus.

(xi) **Vein chlorosis:** Chlorosis will be restricted to tissues adjoining the veins.
Example — Raspberry vein chlorosis virus.

(xii) **Vein clearing:** In systemically infected leaves, a clearing or chlorosis of the tissue in or immediately adjacent to the veins occurs. This pattern is called as vein clearing.
Example — Bhendi vein clearing virus.

(xiii) **Reddening:** Red pigments predominate.
Example — Carrot red leaf virus, clover red leaf virus.

(xiv) **Browning:** Rapid cell death leads to necrosis. Necrosis leads to browning of tissues.
Example — Pea early browning virus.

(xv) **Blackening:** Intensive necrosis leads to blackening of tissues.
Example — Potato black ring spot virus.

(xvi) **Necrosis:** Several types of necrosis are seen in virus infected tissues. Necrosis of specific organs is also common.
Example — Rice necrosis mosaic virus, muskmelon vein necrosis virus, cocoa necrosis virus, sweetclover necrotic mosaic virus.

(xvii) **Wilt:** Plants wilt due to shortage of water or due to impeded supply of nutrients or due to excessive transpiration.
Example — Tomato spotted wilt virus.

(xviii) **Etch:** Desiccation of superficial tissue (epidermal cells) is called etching.
Example — Tobacco etch virus.

(xix) **Leaf rolling, leaf curling, curly top, leaf crinkling, leaf distortion and fern-leaf:** Malformations of leaves are common.
Example — Tobacco leaf curl virus, tomato yellow leaf curl virus, mungbean leaf crinkle virus, strawberry crinkle virus, potato leafroll virus, grapevine fanleaf virus, cotton leaf curl virus, carrot thin leaf virus, beet curly-top virus. Reduced development of the leaf blade in proportion to midrib leads to fern-leaf development.
Example — Asparagus fern-leaf virus.

(xx) **Rugose:** The growth of veinal tissue is retarded.
Example — Citrus leaf rugose virus, Bean rugose mosaic virus.

(xxi) **Enation:** Enations are small outgrowth on leaves, especially on veins or stems.
Example — Pea enation mosaic virus.

(xii) **Big vein:** Veins become big.
Example — Lettuce big vein virus.

(xxiii) **Rosette:** Rosetting is formed due to impeded internodal expansion at stem tips.
Example — Groundnut rosette mosaic virus, peach rosette virus.

(xxiv) **Tumours:** Swellings on stems or roots are also virus disease symptoms.
Example — Sweet clover root tumour virus, cacao swollen shoot virus.

(xxv) **Bud blight:** Necrosis of buds.
Example — Soybean budblight, groundnut bud blight.

(xxvi) **Flower colour-breaking:** Mosaic or variegation of petals of flowers.
Example — Tulip breaking virus.

(xxvii) **Spindle tuber:** Tubers of affected plants become more elongated.
Example — Potato spindle tuber viroid.

(xxviii) **Dwarf, stunt:** Plant growth is very much affected.
Example — Tobacco stunt virus, tomato bushy stunt virus, tomato yellow dwarf virus, sorghum stunt mosaic virus, potato yellow dwarf virus, maize rough dwarf virus.

(xxiv) **Sterility:** Some of the virus affected plants become sterile.
Example — Pigeonpea sterility mosaic virus.

TRANSMISSION OF VIRUS DISEASES

Mechanical Transmission

Some viruses pass directly from plant to plant when their leaves rub together. Mechanical transmission from plant to plant will occur if virus concentration in plant sap and virus stability are high. Tobacco mosaic virus, tomato mosaic virus, cucumber green mottle virus, tobacco necrosis virus and potato virus X are easily transmitted by mechanical contact. The viruses sticking on clothes and hands of workers and equipments are also spread and new infections can occur. Mowing machines spread white clover mosaic and red clover mottle viruses.

Transmission by Grafting

Grafting is a common practice in horticultural crops. Viruses may pass through grafting. Citrus tristeza is easily transmitted by this method.

Transmission by Dodder (Cuscuta)

Dodder is a chlorophyll-less phanerogamic parasitic plant. It grows and entwines its yellowish stem around the stems of its host plants. It produces haustoria, penetrates into the host's vascular bundles and forms natural bridges between diseased and healthy plants. The viruses from diseased plants pass through stems of dodder plant and enter into the healthy plants. Most of the viruses can be transmitted through dodder plants although natural transmission of virus diseases has not yet been reported.

Transmission through Vegetative Propagation Material

Viruses normally invade the host systemically and hence all plant parts contain the virus. Rhizomes, tubers, corms, budwoods and stem cuttings are the common vegetative propagation materials and they may serve as sources for virus infection. Banana bunchy top virus is most transmitted through suckers.

Transmission Through Seed

Several viruses are found in immature seedcoats, consisting of the integuments and nucellar remnants and in perisperm. Since these are part of the mother plant, these parts are infected due to systemic invasion by the viruses. These viruses do not survive desiccation and during maturation of seed coat these viruses perish. There is no vascular contact between embryo and mother plants and hence the viruses cannot move to embryo from seedcoat. However, some viruses like tobacco mosaic virus are highly stable and the seed coat contamination with these viruses may lead to seed transmission. However, these are rare possibilities.

Most seed transmission of viruses is through embryo only. Both embryo and endosperm are formed within the embryo sac after fertilization and lack direct connection and cellular contact through plasmodesmata with the mother plant. Hence virus from mother plant cannot move to the embryo after fertilization. Embryo infection can occur only when mother plant is infected before the production of gametes or before cytoplasmic separation of embryonic tissue from mother plant. Infection after flowering does not lead to transmission through seed or pollen. Viruses that cannot pass beyond the phloem (the viruses which are transmitted by phloem-feeding insects) will not be seed-borne.

Transmission by Insects

(a) Specificity of transmission

Several insects transmit virus diseases. The transfer is not merely mechanical and the relationship between virus and vector is complex. Some easily transmissible viruses by mechanical means (by contact, on hands and equipments) are not transmitted by insects. Mechanically transmitted viruses like tobacco mosaic virus, potato virus X and white clover mosaic virus are not transmitted by insects. Virus transmission by the insects is highly specific. Viruses transmitted by leafhoppers are not transmitted by aphids. Strains of one virus (e.g., barely yellow dwarf virus) may have different insect species for their transmission.

(b) Mouth parts of insects and transmission

Insects have different kinds of mouthparts and ways of feeding. Insects like beetles and caterpillars have biting and chewing mouthparts. They eat away the leaf pieces and damage the plants severely and cells which are touched by their mouthparts are destroyed. Viruses coming through these mouthparts cannot establish on the leaves and hence transmission is rare among the biters and chewers. However, some viruses like turnip yellow mosaic virus and cowpea mosaic virus are transmitted by these beetles.

The majority of the insects that transmit viruses have piercing and sucking mouthparts. Aphids and jassids have this kind of mouthparts. Aphids make initial brief sap-sampling probes between epidermal cells or into them. However, they take up food from phloem vessels only. They need several minutes to hours to penetrate between plant cells rather than through cells to finally reach the phloem. Aphids mostly transmit viruses from and into host parenchyma. Acquisition of the virus and inoculation is in the initial probing period which is of a few seconds to some minutes.

Mouthparts of leafhoppers are most robust compared to aphids. They penetrate the plant cells and reach the phloem within a few minutes. In the phloem, the sap flows under pressure and it will pass up into the insect's food canal. Hence, the leafhoppers transmit the viruses found in the phloem of infected plants.

Thrips have rasping and sucking mouthparts. They feed by gasping open the epidermal cells or by rasping them and sucking up their contents. Thus, thrips can transmit viruses found in epidermal cells also.

(c) Relationship between insect vector and virus

The virus-vector relationship varies widely depending upon the duration of retention of the virus in the vector (persistence). In case of persistent viruses, the virus may simply circulate through the body of the vector or progagate also. Hence, this relationship can be classified as

(i) non-circulative non-persistent,
(ii) non-circulative semi-persistent,
(iii) circulative non-propagative, and
(iv) circulative propagative transmission.

(i) **Non-circulative non-persistent transmission:** Aphids mostly transmit viruses from and into host parenchyma. The virus is acquired and inoculated during brief feeding periods of a few seconds to some minutes, mostly during probes. The aphids become infective immediately after virus uptake.

Persistence in the vector is very brief. In exceptional cases, infectivity persists up to 40 hrs if the vector has no access to plants after acquiring virus. This method of transmission is only mechanical process of stylet contamination. The viruses are 'stylet-borne'. Hence, the transmission is non-circulative non-persistent. The transmission is of low specificity. Aphids which are not pests and not colonizing on the crop, but merely probing surface for a palatable host, can also transmit the viruses efficiently. The aphids such as *Myzus persicae, Aphis craccivora* and *Aphis gossypii* transmit several viruses. However different aphids species may differ considerably in efficiency of transmission.

(ii) **Circulative non-propagative (persistent) transmission:** The viruses taken up by the vector enter the alimentary canal, pass through the gut wall, circulate in body fluid (haemolymph) and contaminate the saliva. The insect becomes infective at the next feeding after a latent period. Infectivity is lost at moulting. Such uptake and inoculation happens only after long feeding times, usually about 15 minutes. The viruses are acquired from the phloem of vascular bundles. Persistent transmission is associated with highly specific-virus-vector relationships. Certain viruses are spread by a single vector species of a biotype, and even individuals may differ in efficiency as vector. Leafhoppers generally transmit viruses in persistent manner.

(iii) **Circulative propagative (persistent) transmission:** Certain viruses are propagated inside the insect body and may last till death. Even transovarial transfer of the viruses through insect eggs is found in leafhoppers, e.g. Rice dwarf, Rice stripe, and Wheat striate mosaic viruses. Such viruses can persist in populations without access to susceptible hosts. Once infective, the insects can transmit the virus to healthy plants throughout their life period.

(iv) **Semi-persistent viruses:** Semi-persistent viruses can be transmitted mechanically also with some difficulty. The viruses are taken up from phloem by the insects along with food. They are adsorbed and gradually eluted from the line of pharynx. The viruses are transmitted through ejection or egestion at the starts of subsequent feeding, e.g., Beet yellows virus.

(v) **Infection of insect vectors by plant viruses:** Harmful effects of some plant viruses on insect vectors have also been reported. The planthopper *Laodelphax striatellus* which transmits rice stripe virus is infected by the virus itself. Individuals from eggs laid by infective planthoppers die prematurely, if they hatched. The mortality of the nymphs is also high, particularly in the first and second instars.

(d) Insects which transmit viruses

Several insects belonging to different orders and families transmit virus. Example for such insects are given below:

Order : Coleoptra

(a) Family: Chrysomelidae (leaf beetles)
Ceratonia spp., *Diabrotica* spp., *Ootheca mutabilis* - Cowpea mosaic virus
Podagria spp. - Bhendi mosaic virus
Psylliodes spp. - Turnip yellow mosaic virus

(b) Family: Coccinellidae (Lady beetles)
Epilachna spp. - Cowpea mosaic virus

(c) Family: Curculionidae (Snout beetles or weevils)
Apion spp., *Sitona lineata* - Broadbean stain virus

(d) Family: Meloidae (Blister beetles)
Epicauta vittata: Bean pod mottle virus

Order: *Homoptera*

(a) Family: Aphididae (aphids)
Aphis craccivora — Groundnut rosette virus
Myzus persicae — Potato leaf roll virus
Toxoptera citricida — Citrus tristeza virus
Cavariella aegopodii — Carrot red leaf virus
Acyrthosiphon pisum — Bean leafroll virus

(b) Family: Cicadellidae (Leafhoppers)
Nephotettix virescens — Rice tungro virus
Cicadulina spp. — Maize streak virus
Circulifer tenellus — Beet curly top

(c) Family: Delphacidae (delphacid planthoppers)
Sogatodes oryzicola — Rice hoja blanca virus
Perkinsiella saccharida — Sugaracane Fiji disease
Delphacodes propingua — Maize rough dwarf virus

(d) Family: Aleurodidae (whiteflies)
Bemisia tabaci — Bhendi vein clearing virus, Tobacco leaf curl virus, Tapioca mosaic virus
Trialeurodes vaporariorum — Beet pseudo-yellows virus

(e) Family: Membracidae (treehoppers)
Micrutalis malleifera — Pseudo curly top virus

(f) Family: Pseudococcidae (mealy bugs)
Planococcus spp. — Cacao swollen shoot

Order : *Thysanoptera*

Family: Thripidae (thrips)
Frankliniella spp. — Tomato spotted wilt
Thrips tabaci — Tomato spotted wilt

Order: *Orthoptera*

Family: Acrididae (grasshoppers)
Melanoplus differentialis: Potato virus X, Tobacco mosaic virus (Mechanical transmission)

Order: *Dermaptera*

Family: Forficulidae (earwigs)
Foficula auricularia — Turnip yellow mosaic

Order: *Heteroptera* (true bugs)

(a) Family: Miridae (plant bugs)
Lygus protensis — Spinach blight virus

(b) Family: Tingidae (lace bugs)
Piesma quardratum — Beet leaf curl virus

(c) Family: Lygaeidae (lygaeid bugs)
Nysius spp. — Centrosema mosaic virus

Order: *Diptera* (flies)

Family: Agromyzidae (leaf miner flies)
Liriomyza langei: Sowbane mosaic virus (Mechanical transmission)

Transmission by Mites

Mites belong to the class Archnida and order Acarina. Mites have puncturing and sucking mouthparts. Mites belonging to two families are known to transmit viruses. Eryophid mites belong to the family, Eriophyidae. *Abacarus hystrix* transmits ryegrass mosaic virus. *Aceria cajani* transmits pigeon-pea sterility mosaic virus. Spider mites belong to the family Tetranychoidae. *Brevipalpus obovatus* transmits citrus leprosis virus while *B. phoenicis* transmits coffee ringspot virus.

Eriophid mites are elongate and tiny, about 0.2 mm long. They are invisible to the naked eye and they often hide between hairs of leaves and in buds. They don't have much movement and are carried by wind. They have slender stylets with which they puncture epidermal cells. The two pads at the tip of the rostrum help in bringing saliva to the stylets and in sucking up plant sap.

Transmission by Nematodes

Some of the nematodes are known to transmit viruses. *Longidorus* and *Xiphinema* species transmit nepoviruses such as the grapevine fan leaf virus. *Trichodorus* and *Paratrichodorus* species transmit tobraviruses like tobacco rattle virus and pea early browning virus.

Nematode-borne viruses are retained in the vector in the lining of the guide-sheath of the odontostyle in *Longidorus* or in the lumen of the odontophore and the oesophagus in *Xiphinema* or in the entire pharynx and oesophagus in *Trichodorus.* Viruses are retained in the nematodes for long periods. Nematodes have life-cycles of about two years.

Transmission by Fungi

Some soil-borne fungal pathogens transmit plant viruses. *Olpidium* spp. (Class: Chytridiomycetes, Order: Chytridiales and Family: Olpidiaceae) transmit cucumber mosaic virus, tobacco necrosis virus and redclover necrotic mosaic virus. *Polymyxa graminis* (Class: Plasmodiophoromycetes, Order: Plasmodiophorales, Family: Plasmodiophoraceae) transmits wheat mosaic virus while *Polymyxa betae* (same family) transmits beet necrotic yellow vein virus. *Spongospora subterranea* (same family, Plasmodiophoraceae) transmits beet necrotic yellow vein virus. All these fungi are pathogens of the host which carry the viruses. The zoospores of the fungi are released from the host and the zoospores carry the virus and transmit it to the susceptible hosts during their infection process. In some cases plant viruses are carried on the outside of the fungi. Examples are tobacco necrosis virus and cucumber mosaic virus. The viruses like lettuce big vein virus are found inside the zoospores. They persist for years in viable resting sporangia. The two types of transmission by fungi can be considered as non-persistent and persistent transmission.

Transmission through Soil and Water

Tobacco mosaic and tomato mosaic virus particles are highly stable and they retain infectivity within soil for long time. They may survive in plant debris also. They infect plants grown in the infested soil.

Similarly, cucumber green mottle mosaic virus survives in water. The transmission of these viruses is only a mechanical process.

SEROLOGY

Alien proteins like that of virus particles can induce the production of specific antibodies in certain cells of lymphatic tissues in animals. The alien protein molecules act as antigens. When cleared blood serum, called antiserum is mixed with the virus suspension precipitates are formed due to specific binding of the virus protein (antigen) with antibodies. The precipitates are visible to the naked eye. The study of such reactions is called serology.

Antisera are prepared by injecting purified virus suspension or suspensions of virus protein intravenously, intramuscularly or both into rabbits or guinea pigs or mice. Normal plant proteins found in the virus suspensions may also induce antibodies. They are removed by saturating the antiserum with sap from healthy plants and then removing the precipitates by centrifuging. The clear antiserum can be stored in refrigerator or deep-freeze or in sealed ampoules after freeze-drying. Antisera have been prepared against a wide range of plant viruses. These antisera are highly useful for detecting virus particles and virus infections for the diagnosis of virus diseases. Several sensitive serology tests have been developed to detect virus infection.

Precipitation Test

The clarified virus suspension (antigen) and antiserum (antibody) are mixed in narrow glass tubes. The precipitates are formed when antigen and antibodies are related.

Ouchterlony Test

Wells are cut in the agar on the bottom of a petri dish. The antiserum (antibody) is placed on the central well while the virus suspension (antigen) is placed on surrounding wells. The antibodies and antigens move towards each other and bands are formed when they meet. The test is also called double diffusion test.

Immuno-Electrophoresis

Antigen (virus) mixtures are first separated in one direction by an electrical current. Then antiserum is put into a long trough at one side of the well parallel to the direction of the current. Antigenic components and antibodies then diffuse together and produce bands. The technique will be useful in distinguishing closely related strains.

Agglutination Test

It is similar to precipitation test. But the virus (antigen) or the antibody is attached to larger inert particles. Large clumps are formed when both virus and antibody react. Thus the antibody-antigen reaction is amplified to facilitate to make observations easier.

Chloroplast-Agglutination Test

Crude sap with the virus is mixed with antiserum on a microscope slide. Chloroplasts clump together with virus in this reaction.

Latex Test

The polystrene latex spheres are first sensitized with antibodies. The sensitized latex can be stored at 4°C for years. By mixing the latex with virus suspension, precipitation is obtained.

ELISA (Enzyme linked Immuno-sorbent assay)

The antibody is first linked with an enzyme like alkaline phosphatase. The virus is then mixed with the antibody enzyme.

The substrate (p-nitrophenyl phosphate for alkaline phosphatase) is degraded by the enzyme action and bright yellow colour appears. The colouration can be measured in a spectrometer. The reaction will be positive only when the virus (antigen) is related to the antibody. The test is extremely sensitive and useful in detecting viruses that occur in low concentration.

Immuno-Fluorescence

Antibodies are labelled with fluorescent compounds such as fluorescin isothiocyanate. When the antibodies combine with viral antigens at the surface of cells or in dissected tissues, fluorescence can be viewed through an optical microscope under ultraviolet light.

Radio Immuno-Assay

Antibodies are labelled with radioactive isotopes as iodine^{-125}. The radioactive isotopes can be viewed through microscope.

ELECTRON MICROSCOPY

Electron microscopes have helped to identify virus particles. They can magnify even objects with a size of 0.2 nm. The smallest virus is tobacco necrosis satellite virus which is 25-30 nm. The caulimoviruses measure 50 nm and reoviruses 65-75 nm. Potex viruses are 470-580 nm long while poty viruses are 680-900 nm.

EPIDEMIOLOGY OF VIRUS DISEASES

Outbreak of virus diseases in a locality depend upon the interplay between virus, host and environment. The factors responsible for the virus disease epidemics are listed here.

1. Introduction of new cultivars and changes in cropping pattern

Virus strains occur in nature. They may differ slightly from each other. They may be differentiated by serology. Particularly they differ in the host range and varietal reaction.

Viruses normally have wide host range. Although viruses can cause heavy damage to the host plants, under natural conditions mutual adaptation of both host and virus occurs. This guarantees maximum survival of both the host and virus. Host selection in natural vegetation is towards resistance and tolerance while virus selection is towards low aggressiveness. Hence wild plants are usually symptomless.

Virus disease outbreaks when new plant genotypes are introduced or when new viruses are introduced from some other place into a place where the plants had not adapted to such viruses. Viruses continue to change by mutation and selective adaptation when passing through their hosts. This may be induced by changes in crop type or cultivar.

This changes in cropping pattern and introduction of new varieties highly favour the outbreak of epidemic of virus diseases.

2. Environmental Factors

Temperature and nutrition play an important role in the development of virus diseases. Tobacco mosaic virus produces local lesions in *Nicotiana glutinosa* below 30°C. But at temperatures above 30°C systemic vascular necrosis and plant death may occur. Potato crinkle mosaic is severe only in cooler periods. The temperature above 23°C suppresses the disease development.

3. Antagonism and Synergism

One virus may drastically alter the predisposition of a host to another one. Increased susceptibility to a second virus by previous infection with another (synergism) has been reported. Potato virus Y increases the severity of infection in potato with potato virus X. 'Double streak' disease of tomato caused by tomato mosaic virus and potato virus X is severely necrotic whereas each virus individually does not cause any necrosis.

Sometimes virus transmission by a vector is dependent on a second virus. This type of synergism is found in rice tungro. The vector, green leafhopper should acquire first rice tungro spherical virus to transmit rice tungro bacillus virus subsequently.

One virus may decrease vulnerability to other virus (antagonism). Mild strains of the virus reduce the symptoms produced by severe strains in case of citrus tristeza.

MANAGEMENT OF VIRUS DISEASES

No effective chemicals are available to control virus diseases. Hence strategies should be developed to manage these diseases effectively. The principles of management of virus diseases are listed below:

1. Reliable disease diagnosis is prerequisite for developing management technology since viruses and even strains may differ fundamentally in their relationships with crops, sources of infection and vectors.
2. Roguing of infected plants in the early stages of crop growth is essential. Each infected plant may act as a source for further spread particularly when vectors are found in the field.
3. Eradication of infected source is essential, if a new virus is found introduced in a locality. To eradicate bunchy top of banana, all the infected plants are removed in Palani hills of Tamil Nadu. Eradication may be difficult to achieve if symptomless hosts, particularly wild plants exist.
4. Weeding is an important operation to contain viruses as several weeds act as symptomless alternate hosts for the virus. Weed control should start before crop emergence because control after establishment of vector populations on them may force the vectors to move on to crops, thus enhancing virus spread. In case of non-persistent viruses removal of weeds upto a distance of about 100 metres from the crop may contain the virus disease.
5. Volunteer plants should be removed. Many virus diseases persist in volunteer plants.
6. Hands, tools and farm equipment should be disinfested with solutions of trisodium phosphate (3-10%) and then with soap. Dipping of hands in skim milk may reduce spread of tomato mosaic virus, cucumber green mottle mosaic virus and potato virus X.
7. Fallowing may reduce or eliminate virus diseases. Example: Onion yellow dwarf virus in onion.
8. Continuous cropping of a single cultivar should be avoided. This may result in the increase in vector population.

9. Growing crops, which are commonly infected by the same virus, in the same field side by side should be avoided. Lucerne is infected by bean leafroll virus and the same virus infects peas, chickpea, cowpea and beans. Hence lucerne should not be grown near the leguminous crops.
10. Vegetative propagation material should be obtained from disease free area. Banana suckers should be free from bunchy top virus. Tapioca cutting should be obtained from tapioca mosaic-free areas.
11. Several viruses are seed-borne. It will be difficult to eradicate virus infection from embryos. Hence virus-free seed should alone be used. Seed health testing centres should certify the pathogen-free seed. For quickly spreading viruses the standard should be severe. For example, to control lettuce mosaic virus in U.S.A. zero infection in 30,000 seeds is being enforced. Mungbean leaf crinkle virus is seed-borne and no seeds should be obtained from the infected field.
12. Basic or foundation seed should be derived from a few virus-free mother plants. Tissue culture is used to free the plants from viruses. Apical meristem is normally free from virus infection. Hence apical meristem is grown in tissue culture medium and the plants are regenerated. The regenerated plants should be tested for freedom from virus and then used.
13. Vector control using pesticides may reduce virus disease incidence. However population densities that are not harmful to crops may disseminate viruses effectively. Particularly the non-persistent aphid-borne virus cannot be controlled by pesticides. The aphids transmit the viruses without colonizing the hosts merely by probing on plants in search for food hosts. Insecticides do not work so much quickly to prevent such spread. However insecticides applied to large areas may reduce infection by non-persistent virus through overall reduction in aphid population. Systemic insecticides will be useful in controlling persistent viruses, because the vectors die before the viruses they have picked up have completed their latency period within them.
14. Destruction of overwintering hosts of vector-insects near cultivated land will reduce virus spread.
15. Cultivation of susceptible crops in soils where soil-borne viruses and their vector have been observed to occur should be avoided.
16. Closer spacing should be adopted to increase crop density. This will decrease initial infestation by insects and obstruct their movement within the crop.
17. Crops should be grown in large fields because air-borne vectors mostly settle along edges of the field and the edge forms a larger proportion of small fields than of larger ones.
18. Open fields should be used for cultivation since windbreaks enhance deposition of air-borne vectors.
19. Early or late planting can be done to avoid the season which favours vector multiplication.
20. Reflective surfaces such as aluminium foil mulch, black plastic mulch, sticky sheets of polyethylene, white polyethylene soil mulch, yellow sticky board traps are useful in reducing vector activity.
21. Pheromones are useful in controlling aphid vector.
22. Taller non-susceptible plant species can be raised as narrow barrier strips or as a mixed crop to intercept alighting aphids. These aphids rid their mouthparts of virus by probes on barrier plants before they reach the crop.
23. Mineral oil can be sprayed on crops. The oil controls non-persistent aphid-borne viruses. The oil interferes with the attachment of virus particles to the vector's mouthparts.
24. Neem oil is an antifeedant. It repels the vectors. Neem oil modifies the feeding behaviour of rice green leafhopper which transmits rice tungro virus. The insect feeds on xylem vessels instead of phloem in the leaves sprayed with neem oil. The virus is present in phloem and not in xylem.
25. Virus inhibitory agents have been isolated from non-host plants. Spraying of these agents induced

resistance against virus diseases in the susceptible plants. Root extract of *Boerhaavia diffusa* and leaf extracts from plants like *Clerodendrum* spp. and *Bougainvillea spectalis* induce resistance in tobacco, tomato, soybean, mungbean and sunnhemp against many viruses. Coconut leaf extract and sorghum leaf extract induce resistance against groundnut bud necrosis caused by tomato spotted wilt virus.

26. Spraying of skim milk induces resistance against tomato mosaic virus in tomato and potato virus X in potato.
27. Mild strains of viruses are used to control severe infection by virulent strains. Citrus tristeza is effectively controlled by this method of cross protection. Tomato mosaic is controlled by inoculating a mild mutant of the virus.
28. Quarantine should be established and it should prevent the movement of virus infected material from state to state.
29. Resistant varieties should be grown to avoid virus infection. Monogenic resistance may break down quickly because of existence of strains of viruses. Durable resistant varieties should be developed.
30. In recent years genetic engineering has been used to develop disease-resistant plants. It will be discused in detail in a later chapter.

REFERENCES

1. Bawden, F.C. 1964. Plant viruses and virus diseases. Ronald Press Co., New York, p. 361
2. Bos, L. 1983. Introduction to plant virology. Longman, London and New York, p. 160
3. Corbett, M.K. and H.D. Sisler. 1964. Plant Virology. University of Florida Press, Gainesville, p. 527
4. Diener, T.O. 1979. Viroid and viroid diseases. John Wiley & Sons, Inc., New York, p. 252
5. Diener, T.O. 1981. Viroids. Scientific American, **244**: 58-65
6. Gibbs, A. and B. Harrison. 1976. Plant Virology. The Principles. Edward Arnold London, p. 292
7. Harris K.F. and K. Maramorosch. 1977. Aphids as virus vectros. Academic Press, New York, p. 559
8. Kado, C.I. and H.O. Agrawal. 1972. Principles and techniques in plant virology. Van Nostrand Reinhold Company, New York, p. 668
9. Kurstak, E. 1981. Handbook of plant virus infections. Comparative diagnosis. Elsevier/North Holland, Amsterdam, p. 943
10. Lamberti, F., C.E. Taylor and J.W. Seinhorst. 1975. Nematode vectors of plant viruses. Plenum Press, London, p. 460
11. Luria, S.E., J.E. Darnell, D. Baltimore and A. Campbell. 1978. General Virology. John Wiley & Sons, New York, p. 579
12. Maramorosch, K. and K.F. Harris. 1981. Plant Diseases and Vectors. Ecology and epidemiolgy. Academic Press, New York, p. 368
13. Maramorosch, K. and H. Koprowski. 1967. Methods in Virology Vol. I, 640 pp. Vol.2 (1967) 682 pp, Vol. 3 (1967) p. 677, Vol. 4 (1968) p 764, Vol. 5 (1971) p 530 Vol. 6 (1972) p 542, Academic Press, New York
14. Martelli, G.P. and M. Russo. 1977. Plant virus inclusion bodies. Advances in virus research, 21: 175-226
15. Martin, S.J. 1978. The biochemistry of viruses. Cambridge University Press, p. 145
16. Matthews, R.E.F. 1981. Plant Virology. Academic Press, New York, p. 897
17. Noordam, D. 1973. Identification of plant viruses. Methods and experiments - Pudoc. Wageningen, p. 207
18. Raychaudhuri, S.P. 1977. A manual of virus diseases of tropical plants. Macmillan Company of India, Delhi, p. 299

19. Raychaudhuri, S.P. and T.K. Nariani. 1977. Virus and mycoplasm a diseases of plants in India. Oxford & IBH Publishing Co., New Delhi, p. 102
20. Smith, K.M. 1972. A text book of plant virus diseases. Longman, London,. p. 684
21. Smith, K.M. 1976. Virus-insect relationships. Longman, London, p. 291.
22. Smith, K.M. 1977. Plant viruses. Chapman and Hall, London, p. 241.
23. Van Regenmortel, M.H.V. 1982. Serology and immunochemistry of plant viruses. Academic Press, New York, p. 268.

5

THE PATHOGENS — MYCOPLASMA-LIKE ORGANISMS, SPIROPLASMAS AND RICKETTSIA — LIKE ORGANISMS

These microorganisms belong to various groups of prokaryotic organisms that are morphologically similar to or identical with members of Mycoplasmatales and Rickettsiales. Their similarity to these two orders has given rise to the terms Mycoplasma-Like Organisms (MLO) and Rickettsia-Like Organisms (RLO). The genus Spiroplasma is in the order Mycoplasmatales.

MLO are elementary, spherical, globular and filamentous bodies. Elementary bodies are 60 to 100 μm in diameter. Spherical bodies are less than 60 μm and the large globular bodies are 150 to 1100 μm in diameter. Inclusion bodies are also seen.

Spiroplasma are helical bodies, measuring 3000 to 25000 nm (3 to 25 μm) in length and 100 to 200 μm in diameter.

RLO are usually rod shaped 200 to 500 μm wide and 1000 to 3000 μm long. They are surrounded by a well defined cell wall and plasma membrane.

MLO

It is difficult to culture these organisms. However, new techniques are being developed to culture them. Phloem sap from palm trees and yucca support MLO. MLO are generally confined to the sieve elements of the phloem and in some cases in the cortical parenchymatous tissues. They are seldom found in xylem or mesophyll cells or in the pericycle of roots. MLO move through sieve pores or plasmodesmata and spread within the plant.

Transmission

Leafhoppers transmit MLO. The acquisition period in leaf-hoppers varies from 3 hour to 5 days. Long incubation periods in the insect body are common. MLO multiply in leafhoppers. The transovarial transmission of MLO is common. MLO pass through a successive number of leafhopper generations.

Mode of Pathogenesis

MLO disrupt normal translocation process in the diseased plant. Cells not associated with the invaded tissue are also adversely affected. Phytotoxins are produced by MLO. Excessive callose formation and degeneration of phloem tissue in MLO diseased plants lead to stunting, wilting, or leaf yellowing. The presence of MLO also disrupts normal growth regulator levels resulting in premature development of adventitious buds and altered apical dominance, leading to proliferation of axillary shoots. This condition is called witches-broom. This imbalance is also responsible for phyllody and antholysis in the flowering stage.

Symptoms of MLO Diseases

Several symptoms associated with virus diseases are also seen in MLO diseases. They are: Witches-broom, phyllody, little leaf, leaf roll, leaf yellowing, dwarfing, stunting, wilt, virescence and decline. Examples: Gingelly phyllody, Brinjal little leaf, Tomato big bud, Sandal spike of sandal wood and Sugarcane grassy shoot.

Management of MLO Diseases

1. Sugarcane setts carry sugarcane grassy shoot MLO. Steam air treatment of sugarcane setts is widely practiced to control the disease.
2. MLO are susceptible to tetracyclines. Tetracycline treatment controls sandal wood tree.
3. The disease is spread by leafhoppers. Hence spraying of insecticides reduces the disease incidence.
4. Treatment with growth regulators like kinetin results in MLO degeneration and plant recovery.
5. Several weeds and wild plants serve as alternate hosts for the MLO. Hence clean cultivation will reduce the disease incidence.
6. Eradication of diseased plants results in reduction in the inoculum.

Spiroplasma

Spiroplasmas are helical prokaryotes lacking a rigid cell wall, but bounded by unit membranes, 8-10 µm in thickness, with RNA granules (ribosomes) and DNA strand in the nuclear region. They are motile and have rotary motility. The only cultured spiroplasma is *Spiroplasma citri*, the causative agent of citrus stubborn disease. Corn stunt disease, yellow dwarf of rice, Aster yellows, pear decline and Peach yellow leaf roll are other examples of spiroplasma diseases.

Citrus Stubborn

The important symptoms are stunting, chlorosis of foliage, loss of apical dominance, poor yield with small misshapen fruits and of poor taste. *Spiroplasma citri* can be cultured and has wide host range. It affects 15 families of wild dicotyledonous plants and 1 of monocotyledonous plants. *S. citri* produces a low molecular weight acidic toxin. The leafhopper *Circulifer tenellus* is a natural carrier (not a transmitting vector) of *S. citri*. The spiroplasma infects the insect and reduces the longevity of the insect. The mode of transmission is not well established. Spiroplasma inoculated by root dipping invades citrus seedlings.

Corn Stunt

The infected plant becomes dwarf, the leaves become chlorotic with distorted margins, the apical dominance is lost and the flowers become small. Leaf hoppers are considered to transmit the disease.

Rice Yellow Dwarf

The affected plants become dwarf and leaves become yellow. Green leaf bug, *Trigopotylus ruficornis* transmits the disease.

Similarity Between MLO and Spiroplasma

Both MLO and spiroplasmas are lacking a rigid cell wall and both are sensitive to tetracyclines. But spiroplasmas are helical in shape and motile. Except for motility spiroplasmas are similar to MLO.

RICKETTSIAE-LIKE ORGANISMS (RLO)

They are small bacteria with a cellular ultrastructure typical of gram-negative bacteria. They have a cell wall unlike MLO and spiroplasmas. Now techniques have been developed to culture RLO. RLO causing Pierce's disease of grapevine, almond leaf scorch and alfalfa dwarf have been transmitted by different mechanical inoculation techniques to their initial hosts where they reproduced typical disease symptoms. RLO produces phytotoxins and these phytotoxins induce characteristic symtpoms of the diseases. Many of the RLO observed in plants are confined either to the xylem or phloem.

Xylem Limited RLO

Pierce's disease of grapevine, Almond leaf scorch, phony disease of peach, alfalfa dwarf, plum leaf scald.

Phloem-limited RLO

Rugose leaf curl of clovers, clover club leaf, little leaf of *Sida cordifolia* , citrus greening, potato leaflet stunt.

Non-tissue Restricted RLO

RLO are also found in parenchyma and meristematic cells. The examples are: Yellows of grapevine, Chlorosis and Aspermy of wheat, Apple proliferation, Carrot proliferation and necrosis of grapevine.

Sensitivity to Antibiotics

Penicillin is effective against RLO. Tetracyclines also control RLO. Sulfa drugs also inhibit RLO.

Mode of Transmission

Leafhoppers (cicadellidae) are the natural vectors of these diseases. A psyllid (*Psyllina* sp.) transmits the citrus greening RLO. A leaf bug, *Piesma quardratum* transmits the RLO of rosette disease of sugarbeet. Soil-borne nematode, *Xiphinema index* transmits yellow disease of grapevine.

Important Diseases Caused by RLO

(a) Pierce's disease of grapevine

The symptom of the disease is similar to vascular wilt diseases. Water conducting system is disrupted. The symptoms include decline of vigor, marginal necrosis or scorching of leaves, decreased production, small fruit and usually death of the plant. Leafhoppers (sharp shooter leafhopper in the sub-family, Tetigellinae or froghoppers in the family Cercopidae) transmit the disease. The RLO is xylem-limited. The RLO is rod shaped and 250 to 500 μm in diameter by 1000 to 4000 μm in length. Weekly foliar sprays of 500 to 1000 ppm oxytetracycline control the disease. Resistant varieties are also available.

(b) Clover Club Leaf

It is phloem-limited RLO disease. The symptoms are similar to yellow type diseases. Yellowing of young leaves, leaf curling and twisting and virescence of flower petals are common symptoms. The RLO is transmitted by a leaf hopper, *Agalliopsis novella.* The RLO multiplies in its vector and the insect remains infective throughout its life. The RLO passes transovarially to the next leafhopper geneation.

The RLO cells are usually 200 to 300 µm in diameter by 1000 to 2000 µm in length. The RLO disease is controlled by spraying penicillin or tetracycline hydrochloride.

Chlorosis and aspermy of wheat and pollen sterility of onion have also been reported to be caused by RLO.

REFERENCES

1. Auger, J.G., T.A. Shalla and C.I. Kado. 1974. Pierce's disease of grapevines: Evidence for a bacterial etiology. Science, 184: 1375-1377
2. Chen, T.A. and C.H. Liao. 1975. Corn stunt spiroplasma: Isolation, cultivation and proof of pathogenicity. Science, **188**: 1015-1017
3. Daniels, M.J. 1983. Machanisms of spiroplasma pathogenicity. Annu. Rev. Phytopathol,. 21: 29-43
4. Davis, R.E. 1974. Spiroplasma in corn stunt-infected individuals of the vector leaf-hopper *Dalbulus maidis*. Plant Dis. Reptr. **58**: 1109-1112
5. Davis, R.E. and R.F. Whitcomb. 1971. Mycoplasmas, rickettsiae and chlamydiae: Possible relation to yellows diseases and other disorders of plants and insects. Annu. Rev. Phytopathol, **9** : 119-154
6. Hampton, R.O. 1972. Mycoplasmas as plant pathogens : Perspectives and Principles. Annu. Rev. Plant Physiol, 23 : 389-418
7. Hopkins, D.L. 1977. Disease caused by leafhopper-borne rickettsia-like bacteria. Annu. Rev. Phytopathol, **17** : 277-294
8. Hopkins, D.L. and H.H. Mollenhauer. 1973. Rickettsia-like bacterium associated with pierce's disease of grapes. Science, **179** : 298-300
9. Maramorosch, K. 1974. Mycoplasmas and rickettsiae in relation to plant diseases. Annu. Rev. Microbiol, **28** : 301-324
10. Maramorosch, K., R.R. Granad and H. Hirumi. 1970. Mycoplasma diseases of plants and insects. Adv. Virus Res., 16: 135-193
11. Nienhaus, F. and R.A. Sikora. 1979. Mycoplasmas, spiroplasmas and rickettsia-like organisms as plant pathogens. Annu. Rev. Phytopathol, **17** : 37-58.

6

THE PATHOGENS — ALGAE

One genus belonging to green algae (Phylum : Chlorophyta; Order: Trentepohliales; Family : Trentepohliaceae), *Cephaleuros* causes diseases in crops. *Cephaleuros parasiticus* causes disease on tea, guava, mango, and citrus, *C. coffeae* on coffee, *C. minimus* on cacao and *C.virescens* on tea. The disease caused by the alga is called 'red rust'.

VEGETATIVE THALLUS

The disc-like thallus of *Cephaleuros* is composed of symmetrically arranged cells which radiate dichotomously from the centre to periphery. The cells are elongated and sometimes barrel-shaped. The thallus cells at the margin are monolayered, while several layers develop in the centre of older discs. Usually the algal pigments extend between the cuticle and the epidermis of host leaves, or more rarely, between the epidermis and palisade cells. Under certain conditions, the alga penetrates between the mesophyll cells of the leaf of the plant. The alga appears to extend by the mechanical force of its expansion rather than by enzymatic dissolution of host structures.

A large airspace is found between the algal disc and the necrotic epidermal cells. Rhizoids arise from the thallus and extend through the airspace to the underlying host cells. The rhizoids serve to anchor the thallus and furnish it with water and other substances from the host.

SEXUAL REPRODUCTION

Large, sessile, flask-shaped cells, known as gametangia arise in the thallus. In the presence of free water they open by fissure and release swarm spores. Eight to 32, but usually 16, biflagellate isogametes per gametangium are found. Copulation takes place between two gametes either from the same gametangium or from different gametangia. The zygote is formed. The zygote produces a dwarf sporophyte, bearing a small dehiscent microsporangium. After meiotic division four quardiflagellate microzoospores are produced. These microzoospores infect the host only rarely.

ASEXUAL REPRODUCTION

The older algal discs produce hair like initials at end of some of the radiating rows of cells. These, after rupturing the cuticle or epidermis, differentiate into either setae or sporangiophores both of which are produced abundantly on the same thallus. The sporangiophores terminate in a swollen apical cell from which many lateral protrusions are subtended. Each protrusion develops into a zoosporangium attached to the apex by a pedicel. Sporangia are disseminated by wind and water. Under favourable conditions sporangia release zoospores through an ostiole. Zoospores are biflagellate and about 30 zoospores are produced from each sporangium (Fig. 41).

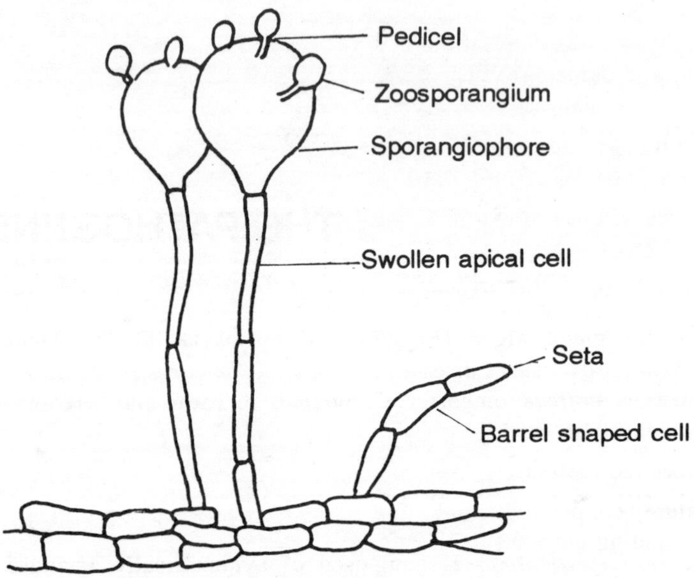

Fig. 41 Asexual structures of *Cephaleuros*

INFECTION PROCESS

When the zoospore reaches the host cell, it comes to rest. Cross-walls are formed repeatedly. Buds develop from the under-surface of primary disc cells and penetrate the host leaf, and form a 'secondary disc' in the plant tissue.

The zoosporangia are disseminated by wind. Rain splashes off the zoospores. Hence at the end of the rainy season the disease becomes severe.

Leaf infection: Yellow-green pinpoint appears in the early stages on the upper leaf surface and rarely on the lower leaf surface. The lamina, veins and petiole are affected. The algal thallus grows mostly subcuticular and the cuticle imparts a glistening appearance to the early developmental stage. The sporophores and sterile hairs contain the orange pigment hematochrome, and the colony becomes velvety in texture and its colour changes from green to orange brown. The thalli become slightly raised. Green islands around the infection foci are common. The cell layers nearest to alga die and assume a cork like appearance. They serve as a barrier to further algal penetration. Only vigorous host is capable of producing an effective barrier. The alga takes up water and minerals from the host by osmosis.

Stem infection: Young stem lesions are normally darkish grey or purple. But after rain they assume the typical rusty red appearance. The bark becomes scaly and cracks. Localized swelling is frequently observed at the site of infection. Lesions coalesce, often girdling the stem completely. Young wood dies, but old wood may recover. Affected branches are stunted and bear fewer leaves which show chlorosis and variegation.

Fruit infection: In guava, citrus, pepper, coffee, tea and avocado, fruit infection is serious. Darkish green to brown or black lesions which are slightly raised and irregular in outline are common on fruits. The alga penetrates only superficial cell layers of the fruit.

DISEASE MANAGEMENT

Host debility is usually accompanied by an increased incidence and severity of the disease. The direct link between host vigour and damage has been reported on citrus, tea, rubber and oil palm. Poor drainage is an important cause for the disease severity.

Application of nitrogen and potassium significantly reduces the disease. Drainage should be provided. Insect, mite and fungal attack increases the disease as they cause general debility.

All badly diseased or dead young wood should be removed. Pruning is recommended for tea, citrus and cacao trees for the control of this disease.

Close plucking of tea induces the formation of thin wood or watershoots. It makes the plant susceptible to the disease and hence it should be discouraged. Shade trees and shelter belts may serve as primary sources of inoculum. Hence shade trees should be selected carefully avoiding the susceptible hosts.

Proper shading of tea and cacao may render the host more resistant to *Cephaleuros.* But close planting of guava and citrus favours the disease as inter-tree competition adversely affects the plant health.

Irrigation reduces red rust due to an increase in host vigour.

Bordeaux mixture, copper oxychloride and cuprous oxide can be sprayed to control the disease. However merely providing nutrients and irrigation, the disease can be effectively managed.

REFERENCES

1. Chowdary, B.K. 1969. Induction of reproductive organs in *Cephaleuros virescens.* Indian J., Microbiol. **8**: 153-158
2. Joubert, J.J. and F.H.J. Rijkenberg. 1971. Parasitic green algae. Annu. Rev. Phytopathol, 9: 45-64
3. Ruehle, G.D. 1941. Algal leaf and fruit spot of guava. Phytopathology, **31** : 95-96
4. Vidhyasekaran, P. and C. Parambaramani. 1971. Carbon metabolism of alga infected plants. Indian Phytopath, **24** : 369-374
5. Vidhyasekaran, P. and C. Parambaramani. 1971. Nitrogen metabolism of alga infected plants. Indian Phytopath, **24**: 500-504
6. Vidhyasekaran, P. and C. Parambaramani. 1972. Mineral metabolism of alga infected plants. Indian Phytopath, **25**: 86-90.

7

THE PATHOGENS - PHANEROGAMIC PARASITES

Some of the plants are parasitic on other plants. They depend upon another vascular plant for food or water. They cause disease by taking away the nutrients of the host. The important phanerogamic parasites are *Striga, Orobanche, Cassytha, Cuscuta* and *Loranthus.*

STRIGA

Many species of *Striga* are known to affect plants. *Striga asiatica* (Syn: *S. lutea, S. hirsuta* and *S. coccinea)* affects sorghum, rice, maize and sugarcane, and *S. densiflora* and *S. euphrasioides* affect sorghum and sugarcane. *Striga* is called witch weed. *Striga* is characterized by opposite leaves and irregular flowers with a pronounced bend in the corolla tube. It is an obligate parasite; it will not develop at all without a host. It is a root parasite and produces haustoria which from a morphological and physiological graft with the roots of the host. It lacks typical root hairs and root caps.

Seed germination: Seeds of *Striga* require after-ripening or post-harvest ripening. After this period 'conditioning' occurs. Certain environmental conditions, involving water and temperature, must prevail before the seeds respond to a chemical stimulant from the host. The conditioning period is about two weeks. After conditioning, seeds germinate in response to a host stimulant and the radicle emerges. Germination is followed by haustorial initiation and development, attachment to the host root and lastly, penetration of the vascular tissue.

For seed germination, *Striga* requires a host root within a few millimeters from seed. Host root exudates induce seed germination. The stimulant in the root exudate has been identified as 'strigol'. Strigol promotes germination of seeds of *S. asiatica* at concentrations as low as 10^{-16} M. Ethylene, cytokinins, gibberellins and coumarins also stimulate germination of *Striga* seeds.

Infection process: *Striga* absorbs water and food stuffs through its haustoria. Carbohydrates and nitrogen of the host are utilised by the parasite. This process results in sugar sink and nitrogen starvation in the host. Differential absorption of phosphorous, potassium, sulphur and iron by the infected host has also been reported. The infected sorghum plants show 90-95% lesser cytokinins and 30-80% lesser gibberellins. A strong transpirational pull from host to parasite is observed. High humidity inhibits growth of *Striga* due to a reduced flow of materials from host to parasite. Due to the same reason *Striga* survives more in dry soils. *Striga* strengthens the roots that are parasitized. It stimulates root production of the host.

The infected plants become stunted with heavy yield reduction. The parasite blossoms and forms fruits within 3 or 4 weeks of emergence, producing several hundred thousand minute seeds per plant. The seeds are dispersed by wind, finally settling on the soil.

Biological strains: Biological strains in *S. asiatica* are common. Sorghum strain does not infect pearl millet and vice versa. Root exudates play an important role in the host specificity. Sorghum exudate does not stimulate germination of seeds of pearlmillet strain and vice versa. However, root exudate alone does not determine specificity. For example, cotton plants produce 'strigol' which stimulates *Striga* seed germination. But cotton is not infected by *Striga.* Haustorial factor is much more important in the pathogenicity of *Striga.*

Control Methods

1. Use of catch crops: Catch crops are hosts that stimulate and support parasitism. Catch crops are ploughed under to prevent parasite maturation.
2. Trap crops: Traps crops stimulate germination but do not support parasitism. They need not be ploughed under. Cotton and groundnut are trap crops for *S. asiatica.* Sudangrass is a trap crop for *S. hermonthica.*
3. Fumigation: Methyl bromide is a fumigant which sterilizes the soil and kills the *Striga* seeds.
4. Solar sterilization: In hot dry areas soil mulching with polyethylene film is useful in controlling-*Striga* seed germination.
5. Weedicide: 2, 4-D and MCPA are highly effective.
6. Hand weeding: The shoots of the parasite should be systematically pulled out before seed set and destroyed.
7. Resistant varieties: Some crop varieties produce less stimulants. *Striga* -resistant varieties have been bred in sorghum and sugarcane.
8. Quarantine: *Striga* is a quarantine object. The imported crop seeds should be free from *Striga* seeds.

OROBANCHE

Orobanche is commonly called as 'broomrape'. Two species of *Orobanche* are common. *Orobanche aegyptica* and *O.cernua* affect tobacco, sunflower, melons, tomato and broadbean. In India tobacco crop is very much affected. Orobanche is also a root parasite. It is an annual fleshy flower plant, growing upto a height of about 12" with cylindrical stem, pale in colour, thickened at base, and covered with scaly leaves that end in spikes. The plant lacks chlorophyll. It produces abundant minute seeds. Seeds remain dormant in soil for many years and when susceptible host is planted, the seeds germinate due to stimulants present in the root exudate. The stimulant has been identified as a benzopyran derivatives. Several growth regulators like ethylene, cytokinins, gibberellins and coumarins also induce seed germination.

Orobanche attacks dicots while *Striga* attacks monocots. In tobacco crop, *Orobanche* emerges 5 to 6 weeks after planting the tobacco and as many as 50 shoots may be found around a single tobacco plant. The affected plants are stunted and chlorosis of the leaves is predominant. The *Orobanche* shoots develop singly or in bunches and they are attached to the host roots in a fragile manner by swollen cloves. They grow quickly, sometimes rising above the height of the host plant.

Carum ajown is the trap crop for *Orobanche cernua* and it can be exploited for the control of *Orobanche.* Insects like *Sciara* sp. eat seedings and stems of *Orobanche.* The insects, *Agromyzid* fly and *Phytophthora orobanchiae* attack *Orobanche.* Hand weeding will be highly beneficial. Resistant varieties have been developed particularly in sunflower.

LORANTHUS

It is a stem parasite on trees. It attacks teak, casuarina, *Acacia,* rubber and kapok. *Loranthus longiflorus* is the common species which attacks sandalwood, rosewood, teak and *Acacia* trees. Custard apple, jack fruit, orange, mango, guava, mulberry, pomegranate, tamarind and jujube are also attacked by *Loranthus.*

Loranthus has green leaves and small berries which are attractive to birds. The birds eat the fruit and the seeds are defected. The defected seeds adhere to the branches as the seeds have sticking properties. The seed coat is gelatinous and able to absorb water from rain, mist or dew. This property prevents the seed from perishing. The natural fall of berries, from higher to lower levels of the same host tree, ensures a continuous infestation. The seeds sprout in due course growing into the host tissue by producing a hold-fast. The hold-fast may become as large as a tennis ball, completely surrounding the branch. The tops of every infected branch die. Very few leaves are formed and losses ranging from 20 to 30 per cent of the harvest have been reported.

For effective control, the parasitized branches should be cut out. Shooting the birds that are known to disseminate the seeds is done in some places. To kill *Loranthus* one hole is made at the bottom of the *Loranthus* and a mixture of 7g of copper sulphate and 1 g of 2, 4-D is applied. Refined diesel oil (Powerine) is also used. Forty per cent emulsion of the diesel oil is prepared in water containing 0.005 per cent washing soap as an emulsifying agent and sprayed on *Loranthus.*

CASSYTHA

Cassytha filiformia is the common species which attacks several plants. It is a slender, thread-like cylindrical herbaceous vine. Leaves are reduced to mere scales. *Cassytha* covers other vegetation like a mantle. It attaches itself to the host by means of haustoria. It resembles dodder but it has a spicy fragrance. It produces fruit which is a globose drupe, 1-2" in diameter and covered by a fleshy receptacle. The birds eat these fleshy fruits and the seeds and distributed by them. Orange trees are very much affected by this parasite. *Cassytha* should be removed manually before it sets seeds.

DODDER (CUSCUTA)

Dodder affects many crops and the important ones are alfalfa and clover. The plant is thread like with leafless stems which are yellow orange or greenish white in colour. Minute leaves which are functionless are also seen in some species. Dodder obtains water and nutrients from the host plant through haustoria which are embedded in the stems of the host plant. A close orientation of the vascular elements of host and pathogen is seen. Some dodders have traces of chlorophyll and there are stomata on the stem.

Dodder plants produce small white flowers and abundant seeds. The seeds fall on the soil and germinate producing tiny colourless seedling which evidences a spiral growth at the upper most portion. These seedlings will survive only for short time if host is not available. When the host is present, it attaches with the host. After the contact between the host and the parasite is established, the portion of the seedling embedded in the soil gradually shrivels and disappears.

Dodder plants should be manually removed to avoid the loss caused by the parasite.

REFERENCES

1. Cordas, D.I. 1973. Effect of branched broomrape of tomatoes in California. Plant Dis. Reptr., **57**: 926-927
2. Eplee, R.E. and M.A. Langston,. 1976. Developments in the control of *Striga* in the U.S.A., PANS, **22**: 61-64

3. Johnson, A.W., G. Rosebery and C. Parker., 1976. A novel approach to *Striga* and *Orobanche* control using synthetic germination stimulants. Weeds Research, **16**: 223-227
4. Kasasian, L. 1971. *Orobanche* spp. PANS, **17**: 35-45
5. King, L.J. 1974. Weeds of the World - Biology and Control. Wiley Estern Pvt. Ltd., New Delhi, p. 526
6. Krishnamurthy, G.V.G., K. Nagarajan and G.H. Chandwani, 1977. Studies on the control of Orobanche on tobacco. Tobacco Research, **2**: 58-62
7. Kuijt, J., 1969. The biology of parasitic flowering plants. Univ. Calif. Press, Berkeley, p. 246
8. Musselman, L.J. 1980. The biology of *Striga*, Orobanche, and other root-parasitic weeds, Annu. Rev. Phytopathol, **18**: 463-489
9. Parker, C. 1965. The *Striga* Problem - A review. PANS, **11**: 99-111
10. Pieterse, A.H. 1979. The broomrape (Oobanchaceae). A review. Abstr. Trop. agric., **5**: 9-35
11. Porwal, B.L and M.K. Mathur, 1972. Control of *Striga* - A review, Indian J. Weed Sci., **4**: 129-135
12. Shaw, W.C., D.R. Shepherd, E.L. Robinson and P.F. Sand. 1962. Advances in Witchweed control. Weeds, **10**: 182-192
13. Verma, J.K. and M.R. Bajpai. 1963. Witchweed in India. FAO Plant Prot. Bull., **11** : 4-7.

8

THE PATHOGENS - NEMATODES

Nematodes are invertebrates. The body of the nematodes is elongate (thread like; nema in Greek means thread) without any segment. It is cylindrical tapering at each end especially towards the tail. The females may swell to become spherical. The size of plant nematodes ranges from 0.2 mm to 10 mm and commonly 0.5 to 1.5 mm range.

The basic body structure of nematodes consists of a flexible body wall which is made up of cuticle, hypodermis and somatic muscles. The wall surrounds a tube-like gut which is made up of oesophagus and intestine. The body cavity between the gut and the body wall is usually regarded as a pseudocoelom containing a pseudocoelomic fluid.

The nerve system consists of a nerve ring which encircles the gut usually in the region of the oesophageal isthmus. Several nerves extend anteriorly and posteriorly.

An excretory duct opens to the exterior via an excretory pore. The gut is an internal tube beginning at the oral opening and ending at the ventrally placed anus in juveniles and females and at the cloaca in males.

Females have a vulva in the mid-body region with paired reproductive tracts. The reproductive tract consists of an ovary, oviduct, uterus, vagina and vulva. Males have single or paired tests, seminal vesicle and vasdeferens. The vasdeferens opens into the cloaca which has a ventrally placed opening. Males usually have a pair of copulatory spicules that lies in an invagination of the cloacal wall. There are sclerotized structures such as a gubernaculum or lateral accessory pieces which guide the spicules (Fig. 42).

The females lay eggs after copulation. The young one from the egg is called larva. The nematodes usually moult four times to reach the adult stage.

PLANT PARASITIC NEMATODES

The plant parasitic nematodes belong to class: Nematodea; subclass: Secernetia (= Phasmidia); Order: Tylenchida; Suborder: Tylenchina and Aphelenchina. The important parasitic nematodes are:

Ditylenchus angustus : Ufra disease of rice

Ditylenchus destructor : Potato tuber rot nematode

Anguina tritici : Wheat ear cockle

Globodera rostochiensis : Potato cyst nematode

Meloidogyne incognita : Root knot nematode of tomato and cucumber

Meloidogyne javanica : Root knot nematode of tobacco, sugarcane, banana and vegetables

Heterodera avenae : 'Molya' disease of wheat and barley

Aphelenchoides besseyi : Rice white tip nematode

Tylenchulus semipenetrans : Citrus nematode causing slow decline of citrus

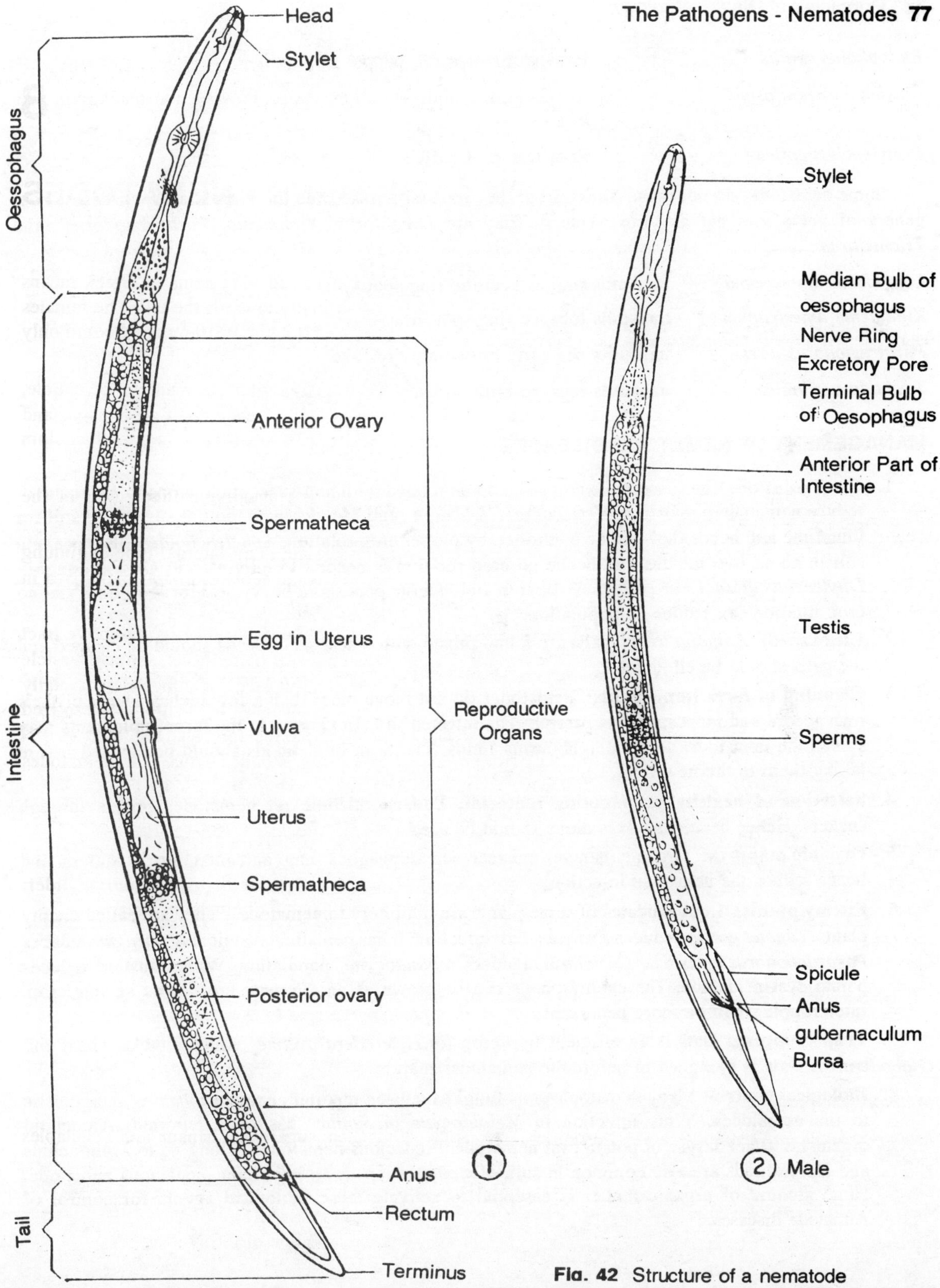

Fig. 42 Structure of a nematode

Radopholus similis	:	Banana rhizoma rot, pepper slow wilt
Rotylenchus reniformis	:	Reniform nematode in cotton, maize, cowpea and blackgram and banana
Pratylenchus coffeae	:	Root lesion of coffee

Some nematodes do not cause diseases but they transmit viruses and the viruses cause diseases. Four genera of nematodes act as virus vectors. They are *Longidorus, Xiphinema, Paratrichodorus* and *Trichodorus.*

Longidorus macrosoma	transmits prunus necrotic ring spot virus
Xiphinema americanum	transmits tobacco ring spot virus
Paratrichodorus teres	transmits pea early browning virus and
Trichodorus similis	transmits tobacco rattle virus.

MANAGEMENT OF NEMATODE DISEASES

1. **Crop rotation:** The susceptible crops should be rotated with non-susceptible crops to eliminate or reduce nematode population. *Heterodera, Globodera* and *Meloidogyne* do not survive in soil for long time and hence they can be controlled by proper crop rotation. *Aphelenchoides* disappears in soil in about two months and hence no crop rotation is needed to eliminate it. On the contrary *Ditylenchus dipsaci* the pathogen of onion and lucerne persists in heavy soil for long time and no crop rotation can reduce its population.
2. **Clean seed:** *Anguina tritici* galls are found mixed with wheat seed. Seeds should be cleaned and the galls should be eliminated.
3. **Cleaning of farm implements:** Nematodes do not move more than a few inches a year of their own accord and most spread is passive. The infested soil sticking on to the farm implements may spread the nematodes to the neighbouring fields. The farm implements should be cleaned before taking them to the new field.
4. **Selection of healthy propagating material:** Banana rhizome rot nematode spreads through suckers. Hence healthy suckers alone should be used.
5. **Organic manures:** Organic manures enhance nematophagous fungi and antagonistic bacteria and hence reduce the nematode infection.
6. **Enemy plants:** Root exudates of some plants are inhibitory to nematodes. They are called enemy plants. *Tagetes erecta* reduces *Pratylenchus* infection. It has nematicidal action. *Asparagus* reduces *Paratrichodorus christiei.* *Crotolaria* reduces *Meloidogyne* population. White mustard reduces potato cyst nematode. The enemy plants can be grown as an alternate crop or as an intercrop. Intercropping will be more beneficial.
7. **Trap cropping:** Oats is an efficient trap crop for *Heterodera avenae* which attacks wheat and barley. Oats is ploughed in before the nematodes mature.
8. **Biological control:** Many nematophagous fungi have been reported. *Pseudomonas* is antagonistic to the nematodes. Virus infection in *Meloidogyne incognita* has been reported. Amoeboid organisms attack larvae of potato cyst nematode. Predacious nematodes belonging to Mononchids and Dorylaimids are also common in soil. *Mononchus* sp. is an important predacious nematode. Huge amount of organic matter is essential to activate these biological agents for control of nematode diseases.

9. **Physical methods:** The infested soil can be sterilized with steam. Potato cyst nematodes are killed by passing the steam into the soil through perforated pipes in western countries. Hot water treatment will be beneficial to control cyst nematodes in potato tubers. Rice white tip nematode can be eliminated by hot water treatment.

10. **Chemical control:** Many fumigants are available to eradicate nematodes. Methyl bromide (Dowfume MC-2) at 400-900 kg/ha, dichloropropene (Telone) at 500-900 kg/ha, dibromochloropropane (Nemagon) at 120 kg/ha and metham - sodium (Vapam) at 1200 1 solution per ha are injected into the soil to eradicate nematodes.

    Some of the granules also reduce nematode population. Aldicarb (Temik) at 11 kg/ha and oxamyl (Vydate) at 11 kg/ha effectively control nematode. Carbofuran 0.75-1.0 kg/ha also reduces nematodes.

    Seed treatment with chemicals is also beneficial. Fumigation with methyl bromide controls stem nematode *(Ditylenchus dipsaci)* which is carried on onion and lucerne seeds. Sodium hypochlorite solution containing 1% available chlorine destroys the cysts of *Globodera rostochinensis,* the potato cyst nematode found contaminated with potato tubers.

11. Resistant varieties have been developed against many nematodes.

12. **Quarantine:** Potato cyst nematode is a quarantine objective and strict quarantine prevents the movement of the nematode from one country to another.

REFERENCES

1. Allen, M.W. and S.A. Sher. 1967. Taxonomic problems concerning the phytoparasitic nematode. Annu. Rev. Phytopathol, 5: 247-264
2. Bird, A.F. 1981. The structure of nematodes. Academic Press, London and New York, p. 318
3. Bird, A.A. 1974. Plant response to root-knot nematode. Annu. Rev. Phytopathol, 12: 69-85
4. Croll, N.A. 1970. The behaviour of nematodes. Arnold Press, London, p. 117
5. Southey, J.F. 1982. Plant Nematology. Her Majesty's Stationery Office, London, p. 440
6. Webster, J.M. 1972. Economic nematology. Academic Press, London and New York, p. 563
7. Zuckerman, B.M., W.F. Mai and R.A. Rohde. 1970. Plant Parasitic Nematodes. Vols. I and II. Academic Press, London and New York, pp. 345 and 347 respectively.

9

NON-PATHOGENIC DISEASES

Besides pathogens, some physiological disorders also result in diseases. These diseases are called non-pathogenic diseases or physiogenic diseases. Most of these diseases are caused by deficiency in nutrition.

NITROGEN DEFICIENCY

Nitrogen deficiency is characterized by overall chlorosis of the older leaves. In severe cases the whole plant is stunted and leaves remain small. During periods of high nitrogen demand when growth rates are high, cereal crops commonly show yellowing of the leaf tips. Lack of nitrogen uptake results in earlier maturity of many crops. Nitrogen deficiency can be corrected by applying nitrogenous fertilizer in split doses. Urea 2% can be sprayed and the spray should be repeated at least 2 times. Diammonium phosphate can also be sprayed.

PHOSPHORUS DEFICIENCY

Crops deficient in phosphorous show an overall stunting in early growth. Reddening or purpling of older leaves and stems due to increased anthocyanin production is an important symptom. Foliar application of phosphorus in the form of diammonium phosphate can be recommended.

POTASSIUM DEFICIENCY

First symptoms are seen on the older leaves. Initial marginal chlorosis spreads towards the main vein. Bleaching or chlorosis of the leaf tip is common. The plants become stunted with loss of turger. The plants become less drought resistant and wilt under high evaporation conditions. Potassium sulphate can be applied to soil or sprayed on the foliage.

MAGNESIUM DEFICIENCY

Interveinal chlorosis with a broad veinal area remaining green is the typical symptom of the magnesium deficiency. Intense anthocyanin purpling is common. In severe deficiencies the chlorotic areas turn necrotic and may drop out leaving holes in the leaf. The general symptom expression is caused by disruption of chloroplasts and loss of chlorophyll resulting in the initial chlorosis. High potassium levels generally cause a reduced uptake of magnesium.

Potassium: Magnesium ratio is important. Foliar application of magnesium sulphate may rectify the deficiency. But only young crop will respond to foliar application and old crop will not respond. Many crops show magnesium deficiency in the older leaves as senescence takes place.

In cotton magnesium deficiency in the old crop is common. Anthocyanin purpling is very commonly seen on old leaves.

SULPHUR DEFICIENCY

Chlorosis due to sulphur deficiency resembles nitrogen deficiency except that symptoms generally appear in the younger leaves first. In nitrogen-deficient plants symptoms appear first in the older leaves. Sulphur deficiency can be rectified by applying sulphur containing fertilizers such as ammonium sulphate (24%), super phosphate (11.14%), potassium sulphate (18%) and gypsum (14.18%). Foliar application of magnesium sulphate may also reduce the symptoms.

IRON DEFICIENCY

Iron deficiency is common in calcareous soils. It is due to presence of bicarbonate ion. This ion reduces uptake and translocation of ion in the plant. Inundation of water and heavy rains cause iron deficiency. High levels of copper, nickel or zinc can induce iron deficiency. In sewage sludge applied field also, iron deficiency is noted. Plants take up ferrous iron or as iron chelates. Reduction of ferric ion to ferrous ion takes place in the rhizosphere. In calcareous soils or poorly aerated soils such reduction does not take place.

The important symptom is general chlorosis of the younger leaves with the veins remaining green. Young leaves may turn completely white followed ultimately by tip die back. This type of iron deficiency is commonly seen in sugarcane crop.

Soil application of iron will not rectify the symptoms as soil-applied iron will get fixed and may not be available. Iron chelate can be applied. Fe-EDTA at 1 kg/ha can be sprayed. Several applications at 14 days intervals may be needed.

MANGANESE DEFICIENCY

Over liming leads to manganese deficiency. In sandy soils manganese deficiency can occur. In soils with high pH also manganese deficiency is noticed. Waterlogging increases availability of manganese and hence in rice fields, deficiency of manganese may not occur. The deficiency symptoms are interveinal yellowing and green band along the vein. The leaf tip remains green, even if the leaf breaks in the middle and hangs down. The deficiency can be corrected by spraying manganese sulphate at 8-9 kg/ha in 250 litres of water. A wetter like teepol should be added.

BORON DEFICIENCY

Due to boron deficiency, the growing point dies stimulating growth of lateral buds. Petioles become necrotic and leaves become chlorotic and brittle. Borax (20 kg/ha) should be applied before sowing to correct the deficiency. Soluble borax (Solubor) can be sprayed at 5 kg/ha.

ZINC DEFICIENCY

Zinc deficiency is very common in rice. Interveinal chlorosis is the common symptom. Necrotic spots also appear. Zinc deficiency in rice can be corrected by spraying zinc sulphate at 0.5% (500 litres/ ha). The spraying should be repeated 2 to 3 times at 5 days interval. Urea at 1.0% mixed with zinc sulphate increases the efficacy of zinc sulphate.

COPPER DEFICIENCY

Copper is not very mobile within the plant. Copper is concentrated in the plant roots and little is translocated to the shoot. Due to copper deficiency yellowing of the tips of the youngest leaf is seen. It is often followed by spiralling of the leaves. Earheads are trapped in the leaf sheath and those that emerge have white tips that are not filled with grain. Soil application of copper sulphate at 30-60 kg/ha is recommended to control the deficiency. Foliar spray of copper oxychloride fungicides at 2 kg/ha may reduce the deficiency syndrome.

MOLYBDENUM DEFICIENCY

Whiptail symptoms are the important syndrome of molybdenum deficiency. The growth of leaf lamina is severely restricted giving narrow strap-like leaves. The remaining lamina is irregular in outline and puckered. Young plants show yellowing and inward cupping of the leaves. Molybdenum deficiency is common in cabbage and cauliflower. Foliar application of ammonium or sodium molybdate at 0.25 g/litre of water may correct the deficiency syndrome.

FOLIAR APPLICATION OF NUTRIENTS

Crop may not be able to take up more than 5 kg/ha of a particular element from one foliar spray application. But the crop requires macronutrients in large quantities. Hence micronutrients alone can be applied as foliar spray effectively. Nitrates and chlorides are not taken up by the foliage and phytotoxicity is common. Hence, they should be avoided. It is better to add a wetter like teepol to increase the efficacy of the foliar spray. Repeated spraying at shorter intervals may be necessary as leaves do not absorb nutrients effectively unlike roots.

REFERENCES

1. Archer, J. 1985. Crop nutrition and fertiliser use. Farming Press Ltd., Suffolk, p. 258
2. Bear, F.E. and R. Coleman. 1949. Hunger signs in crops. Amer. Soc. Agron. And Natl. Fertil. Assoc. Washington, D.C.

10

PHYSIOLOGY OF PLANT PATHOGENS

Fungal and bacterial pathogens obtain their nutrition from host cells. They depend upon the host for the synthesis of various constituents of their cells. The fungal and bacterial cell consists of chitin, cellulose, glucose, mannans, uronic acids, rhamnose, xylose proteins, lignin, lipids, nucleic acids and pigments. The cell constituents are made up of carbon (50-60%), nitrogen (1-10%), phosphorous, sulphur, potassium, magnesium, iron, zinc, manganese, copper, molybdenum, calcium, scandium, vanadium and gallium. In short, the fungal and bacterial cell is made up of carbohydrates, fats, proteins, nucleic acids and minerals. The host plants supply carbon, nitrogen and minerals from which the fungi and bacteria themselves synthesize all the complex cell structures for their growth. The synthetic process requires energy and the fungi and bacteria obtain energy by biological oxidation.

Biological Oxidation

Energy from sunlight is converted in the green plant to chemical bound energy and the energy inherent in these bonds can be released when they are broken. This is due to the removal of electrons from the sphere of attraction of the atom being oxidized and their passage into that of the oxidant. This 'electron transfer' releases free energy. For example when acetaldehyde is oxidized to acetic acid, approximately 61,000 calories are released. The chemical reactions through which energy is made available to the cell are called biological oxidation. The biological oxidation is otherwise called 'respiration'.

Storage of Energy

During the oxidation process, lot of heat energy is released. If the released energy is not utilised, it will go as a waste. Hence the released energy is trapped in a storage battery-like mechanism. Phosphorus plays an important role in this. High energy phosphate bonds are present in co-enzymes. These phosphate bonds trap the energy and store it in their bonds. Whenever energy is required, the energy stored in co-enzymes is released. Thus co-enzymes serve as a battery cell in releasing energy whenever required.

Enzymes

The various chemical reactions through which energy is released to satisfy the requirement of cells for formation and maintenance are activated by various enzymes. Enzymes are defined as complex proteins which catalyze reactions generally essential to the activities of the cell.

Co-enzymes are attached to the enzymes but they are readily dissociable. Enzymes release the energy while co-enzymes trap the released energy. The important co-enzymes are Adenosine Triphosphate (ATP),

Adenosine Diphosphate (ADP), Nicotinamide Adenine Dinucleotide Phosphate (NADP) and Nicotinamide Adenine Dinucleotide (NAD).

Both ATP and NADP are triphosphates while ADP and NAD are diphosphates. We have seen that the energy is trapped by phosphate bond and the trapped energy is released when phosphate bond is broken. In this process ADP is converted into ATP by the addition of one phosphorous bond and energy is trapped ATP is converted into ADP when energy is released. Similarly, NAD is converted into NADP and NADP into NAD.

Redox System

The electron transfers occur along with hydrogen atom. During the respiration process the co-enzymes get reduced by accepting electrons from hydrogen. NADP becomes $NADP^+$ when it accepts hydrogen. This hydrogen and electrons are transferred to oxygen through various steps passing through a system called 'Reduction-oxidation potential (Redox) system'.

Some substances are having chemical bonds exhibiting a stronger electron-accepting tendency while others can give up electrons more readily. When electrons are given up it is called 'oxidation' and when electrons are taken up it is called 'reduction'. Several chemicals take part in the oxidation-reduction process in the fungal cells. For example, if electrons are transferred directly from hydrogen to oxygen 56,580 colories of heat energy is released. If electrons from hydrogen are transferred to cytochrome C only 31,670 colories of energy is released and when electrons from cytochrome C are transferred to oxygen, the remaining 24,910 calories of energy is released.

$H^+ \longrightarrow$	O_2^-	=	56, 580 calories
$H^+ \longrightarrow$	Cytochrome C	=	31, 670 calories
	or		
Cytochrome $C^+ \longrightarrow$	O_2^-	=	24, 910 calories
	Total	=	56,580 calories

Instead of transferring electrons from hydrogen to oxygen directly, if they are transferred through cytochrome C to oxygen, the energy is released in instalments. This prevents wastage of heat energy. In the biological system, several 'Redox' systems exist, which release the energy in 'instalments'. Thus 'Redox' systems are useful in the conservation of energy.

Energy is released through different pathways in fungal and bacterial cells. The important ones are Embden-Meyerhof (E.M.) pathway, pentose phosphate cycle, and Kreb's cycle.

E.M. Pathway

Glucose and sucrose are synthesized in plants through photosynthesis. Glucose is utilized for the energy synthesis. Sucrose is converted into glucose and fructose and glucose and fructose are inter-convertible. Glucose is the starting point for the various oxidation processes.

Glucose

↓

Glucose-6 Phosphate

↓

Fructose - 6 Phosphate
↓
Fructose - 1,6 Diphosphate
↓
Dihydroxy Acetone Phosphate
+
3-Phospho Glyceraldehyde
↓
1,3-Diphosphoglyceric Acid
↓
3-Phosphoglyceric Acid
↓
2-Phosphoglyceric Acid
↓
2-Phosphoenol Pyruvic Acid
↓
Pyruvic Acid

Pentose Phosphate Cycle (Hexose Monophosphate Pathway)

Glucose → Glucose-6-phosphate → 6-phospho-glucono lactone
↓
3-Keto-6-P-gluconic acid
↓
Ribulose-5-phosphate
↓
Sedoheptulose7-P + Glyceraldehyde-3-P
↓
Fructose-6-P + Erythrose 4-P

Kreb's cycle (Citric acid cycle)

Glucose (E.M. Pathway)
↓
Pyruvic acid
↓
Acetyl CO-A
↓
Oxalacetic acid → Citric acid
↓
Cis-aconitic acid
↓
Isocitric acid
↓
Oxalosuccinic acid
↓
α-Ketoglutaric acid → Succinic acid
↑
Fumaric acid
↑
Malic acid
↑
Oxalacetic acid

Terminal Respiration (Oxidative Phosphorylation)

In all the three respiratory (biological oxidation) pathways hydrogen along with electrons are released from different chemicals. These electrons are transferred to oxygen through Redox systems. The important Redox systems involved are cytochromes a, b and c. Cytochrome oxidase and cytochrome reductase are the important enzymes and NAD and NADP are the important co-enzymes which carry the electrons and hydrogen. The energy released is trapped in the high-energy phosphate bonds of co-enzymes. This process is called *Oxidative Phosphorylation.*

Instead of cytochromes, ascorbic acid and phenols are also involved, particularly under stress conditions. Ascorbic acid is oxidized to dehydroascorbic acid and phenols are oxidized to quinones and thus ascorbic acid - dehydroascorbic acid and phenols-quinones are the important 'Redox System'. The enzymes involved are ascorbic acid oxidase and polyphenol oxidase respectively.

When H_2 is transferred to O_2, hydrogen peroxide is formed

$$H_2 + O_2 \longrightarrow H_2O_2$$

Hydrogen peroxide is toxic to the fungi. Hence, it is converted to water (H_2O) by any one of the two enzymes, peroxidase and catalase. Peroxidase removes one atom and transers to another chemical for oxidation. For example, peroxidase can oxidize phenols to quinones. Catalase cleaves and breaks down H_2O_2 to H_2O.

Synthesis of Carbohydrates

We have seen how energy is synthesized in the fungal and bacterial cell. This energy is used for building up of the fungal and bacterial body inside the host tissue. The fungi and bacteria synthesize polysaccharides from simple sugars like glucose and fructose which are synthesized by the plants by photosynthetic process. Polysaccharides may be synthesized either by phosphorylation or by transglycosidation.

Phosphorylation

Glucose is converted into glucose-1-phosphate. When glucose is activated by addition of phosphorous it is converted as follows:

$$\text{n Glucose - 1 - P} \longrightarrow \text{Polysaccharide + Phosphate}$$

The length of the polysaccharide chain will go on increasing by the addition of phosphorylated glucose molecule.

$$\underset{\text{Polysaccharide}}{(C_6H_{10}O_5)\ x} + \underset{\text{Glucose - 1-P}}{C_6H_{11}O_5 - P} \longrightarrow \underset{\text{Polysaccharide}}{(C_6H_{10}O_5)\ x} + P$$

Transglycosidation

Two molecules of glucose will unite to form a disaccharide. Many disaccharide molecules may unite to form a complex polysaccharide and number of monosaccharides.

Transglycosidase transfers glycosyl moiety to another molecule.

$$\underset{\text{Disaccharide}}{nC_{12}H_{22}O_{11}} \longrightarrow \underset{\text{Polysaccharide}}{(C_6H_{10}O_5)\ n} + \underset{\text{Monosaccharide}}{(n)\ C_6H_{12}O_6}$$

Synthesis of Nucleic Acids

DNA

The fungal and bacterial nucleus contains chromosomes which consist of nucleoproteins (proteins plus nucleic acids). They are the carriers of genes (the factors that control inherited traits). Genes are nucleic acids and more specifically deoxyribonucleic acids (DNA). DNA molecules consist of a sugar (deoxyribose), two purines (adenine and guanine), two pyrimidines (cytosine and thymine), and phosphorus. DNA molecules consist of two polynucleotide chains twisted together in a twin helix (Fig. 43) with the bases (purines and pyrimidines) towards the inside and the sugar phosphate backbones on the outside. The two chains are held together by hydrogen bonds between specific pairs of bases, the adenine of one chain being paired with thymine of the other and guanine being paired to cytosine (Fig. 44). The DNA molecules are extremely large and contain more than 2,00,000 base pairs.

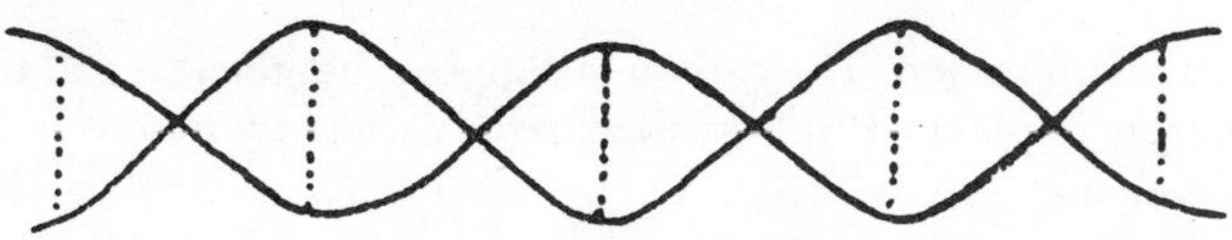

Fig. 43 Helix of DNA

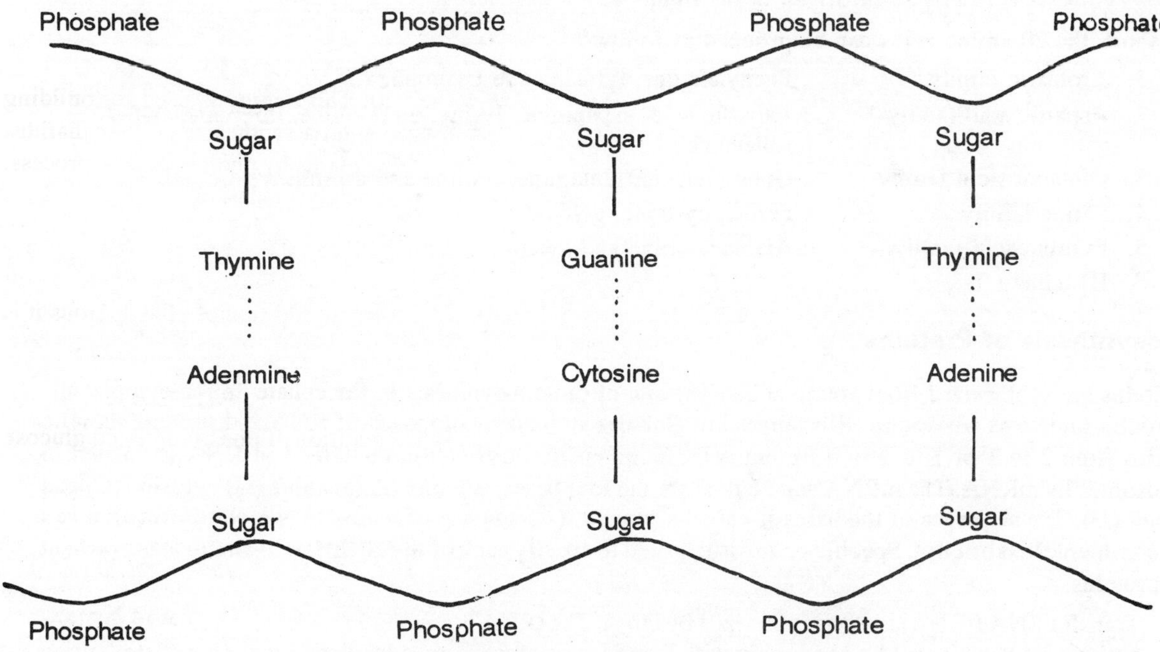

Fig. 44 Structure of DNA

Replication of DNA comes about first by separation of the two strands. Each of the two strands acts as a template (the molecular structure or a compound that serves as a pattern for the production of another compound). It results in the formation of two replicas of the original double-stranded molecule. The order of the bases is not regular and the precise sequence of the bases determines the genetic information carried by the DNA molecule

RNA

Ribonucleic acid (RNA) is manufactured in the nucleus by DNA. RNA consists of a sugar (ribose), two purines (adenine and guanine) and two pyrimidines (cytosine and uracil). It is a single stranded chain. The information contained in the genes is then carried to the cytoplasm by RNA. The RNA carries the genetic code to the site of protein synthesis, and the RNA carrying the message from DNA is called messenger RNA (mRNA).

Two more kinds of RNA are found in fungal cell. They are

(i) transfer RNA (tRNA) which transfers amino-acids to ribosomes where protein synthesis occurs, and

(ii) ribosome RNA (rRNA) which performs protein synthesis.

Biosynthesis of Amino Acids

Amino acids contain carbon and nitrogen. The carbon is supplied by glucose and the nitrogen is supplied by ammoniaeal nitrogen present in plants. Plants synthesize its carbon by photosynthetic process and takes up ammoniacal nitrogen from soil.

Twenty amino acids are commonly present in fungal and bacterial cell. These amino acids are synthesized through E.M. Pathway, pentose phosphate pathway and Kreb's cycle. The pathway of amino acids synthesis is briefly summarized in the figure 45.

In short, the 20 amino acids can be grouped as follows:

1. Aromatic family : Phenylalanine, tyrosine and tryptophan
2. Aspartic acid family : Aspartic acid, asparagine, lysine, methionine, threonine and isoleucine
3. Glutamic acid family : Glutamic acid, glutamine, proline and arginine
4. Serine family : Serine, cysteine, glycine
5. Pyruvic acid family : Alanine, valine and leucine
6. Histidine

Biosynthesis of Proteins

Proteins are synthesized from amino acids. The site of protein synthesis in the cells is in the cytoplasmic particles known as ribosomes. Ribosomes are globular structures composed of RNA and protein, its ratio varies from 2 to 1 or 1 to 1 in different cells. The genetic information contained in DNA is carried to ribosomes by mRNA. The mRNA code letters are the four bases, adenine (A) guanine (G) cytosine (C) and uracil (U). The sequence of the bases is called a 'codon' (a sequence of letters by which information can be communicated briefly). Specific codon is required to specify each of the 20 different amino acids present in proteins.

Transfer RNA (tRNA) is also synthesized by DNA. The tRNA is smaller in molecular size than mRNA. The amino acid is activated by combining with P and forms amino acid adenosine monophosphate (amino acid AMP). Each amino acid attached to AMP is transferred to a tRNA. The tRNA carries as anticodon (counter-acting the code) that can form a temporary bond with one of the codons in mRNA. The tRNA reads the message of mRNA and delivers the amino acid to its appropriate place so that a defined protein sequence can result. The ribosomes contain surfaces that orient amino acid - tRNA units and mRNA molecules so that proper sequences are followed. The ribosome directs the formation of polypeptide chains to carry further synthesis.

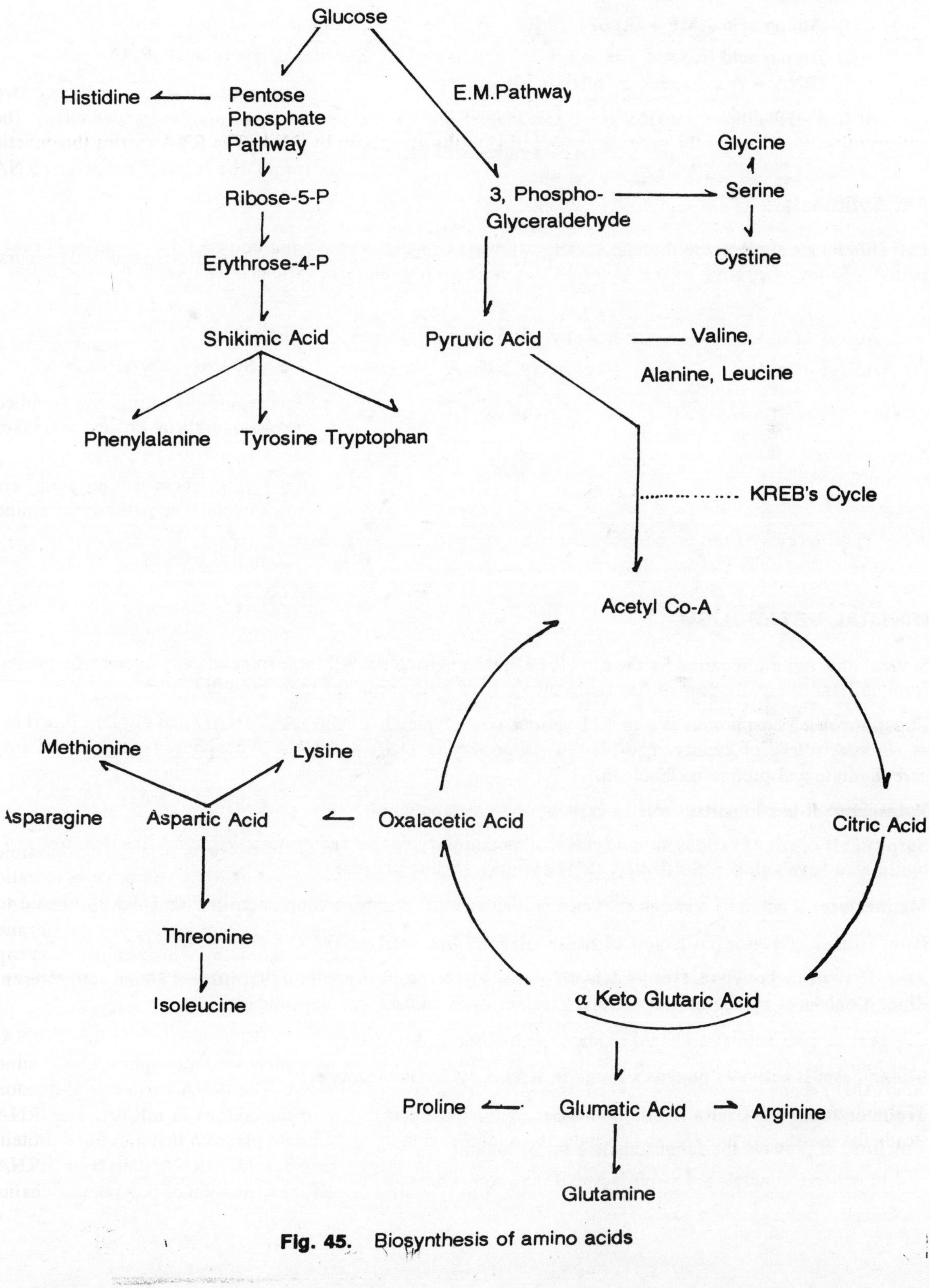

Fig. 45. Biosynthesis of amino acids

(i)	Amino acid AMP + tRNA	→	Amino acid tRNA + AMP
(ii)	Amino acid tRNA + amino acid tRNA + etc. directed by mRNA Ribosomes	→	Specific polypeptide + tRNA
(iii)	Polypeptide + Polypeptide + Polypeptide	→	Protein

Biosynthesis of proteins

Fat Synthesis

Fats (lipids) are synthesized through acetate pathway. Glucose is converted into acetyl-CoA through E.M. pathway followed by Kreb's cycle. Acetyl-Co A proceeds to synthesize various fatty acids as detailed below:

Acetyl - Co A ⟶ Malonyl - Co A ⟶ Palmitate

Acetyl - Co A ⟶ Acetoacetyl - Co A ⟶ P - hydroxybutyryl - Co A

↓

Crotonyl Co A

↓

Butyryl - Co A

↓

Palmityl - Co A

MINERAL METABOLISM

Several minerals are required for the growth of fungi and bacteria. All these minerals are taken up by plants from soil and the pathogens utilize these minerals of host origin for their growth.

Phosphorous: Phosphorous is a part of various co-enzymes like ADP, ATP, NAD and NADP. It serves as storage battery of energy. Phosphorous is present in DNA and RNA. Phosphorous is required for carbohydrate and protein metabolism.

Potassium: It has important role in carbohydrate metabolism.

Sulphur: It is part of various amino acids like methionine, cystine and cystein. Vitamins like thiamine and biotin also have sulphur. Sulfhydryl (SH) enzymes require sulphur.

Magnesium: It activates various enzymes particularly the enzymes which activate transphosphorylation.

Iron: It is an activator (co-factor) of many enzymes like catalase and cytochrome oxidase.

Zinc: Pyruvic carboxylase, ethanol dehydrogenase and tryptophan synthesizing enzyme are all activated by Zinc. It enhances cytochrome synthesis. Zinc activates enolase and depeptidase.

Copper: Copper activates phenol oxidase, peroxidase and tyrosinase.

Manganese: It activates enzymes found in Kreb's cycle and arginase.

Molybdenum: It activates nitrate reductase.

Calcium: It protects the fungus against metal toxicity.

In general minerals act as co-factors for various enzymes.

VITAMINS

The fungus requires minute amounts of certain organic compounds in addition to those which yield energy or are used for structural purposes. Those organic compounds are called vitamins or growth factors. The fungus can synthesize the vitamins by itself. But some of the fungi are deficient for some vitamins. For example, *Phytophthora* spp. are deficient for thiamine. *Pyricularia oryzae* is deficient for biotin. *Ceratostomella ulmi* is deficient for pyridoxine. Plants supply these vitamins to the fungi.

Vitamins are part of co-enzymes.

Thiamine: Thiamine pyrophosphate is the co-enzyme of carboxylase. Carboxylase converts pyruvic acid into carbon dioxide and acetaldehyde and in the absence of thiamine, pyruvic acid accumulates.

Biotin: Biotin is the respiratory co-enzyme (co-enzyme R). Biotin plays an important role in the synthesis of aspartic acid. Assimilation of NH_3 does not take place unless biotin is present.

Nicotinic acid: Nicotinamide adenine dinucleotide (NAD) and Nicotinamide adenine dinucleotide phosphate (NADP) are the important co-enzymes containing nicotinic acid. These two co-enzymes are involved in terminal respiration (oxidative phosphorylation).

Pantothenic acid: It forms a part of the co-enzyme A. Co enzyme A takes part in Kreb's cycle and fat metabolism.

Pyridoxin: It is a part of the co-enzyme, pyridoxal phosphate. This co-enzyme takes part in the transformation of tryptophan into indole.

Thus, the fungal and bacterial pathogens take up nutrients from the plant and synthesize their body.

REFERENCES

1. Bisset, K.A. 1962. Bacteria. E and S Livingston Ltd., Edinburgh and London
2. Gunsalus, I.C. and R.Y. Stainer. 1960. The bacteria. Vol. I. Academic Press, Inc., New York
3. Pelczar, M.J. Jr. and R.D. Reid. 1965. Microbiology. McGraw Hill Book Co., New York, p. 662.

11

PHYSIOLOGY OF PATHOGENESIS

Pathogens are different from parasites. Parasites penetrate into the host and obtain nutrients from the host. They live inside the host without causing any other damage. They may be sometimes symbionts also i.e. they may be beneficial to the host. The examples for parasites are *Rhizobium* spp. which form nodules in leguminous plants and mycorrhizal fungi. Pathogens not only take nutrients from the host but also damage the host severely. They induce several symptoms of the disease and even kill the plants.

Pathogens in their life cycle enter into two distinct phases. One is 'biotrophic' and the other one is 'necrotrophic' phase. In the biotrophic phase they invade the host cells inter - and intracellularly and establish themselves in the living cells. In the second phase, they kill the cells and obtain nutrition from the dying cells. This phase is called necrotrophic phase. In case of obligate pathogens (the pathogens which can live only on living cells) the biotrophic phase may be much longer while in case of facultative saprophytes (normally pathogenic but can live as a saprophyte also), long biotrophic phase may be followed by necrotrophic phase. In case of facultative parasites (normally saprophytes but can live as a pathogen also) necrotrophic phase will be much longer with short biotrophic phase. However in case of all pathogens, both the phases may exist either as a short or long spell.

In the biotrophic phase the pathogens establish themselves in the host cells. The first barrier between the host plant and the attacking pathogen is the cuticle which consists of a structural polymer, cutin, impregnated with wax. Cutin is a polyester composed of hydroxy and epoxy fatty acids. Suberin is another protective barrier on the surface of plants. It is also a lipopolymer. Host cells have cell wall, a structure surrounding the protoplast. Cell walls contain complex polysaccharides and proteins which act as barrier for entry of the pathogens. Cell wall is divided into three regions:

(i) middle lamella,
(ii) primary wall, and
(iii) secondary wall.

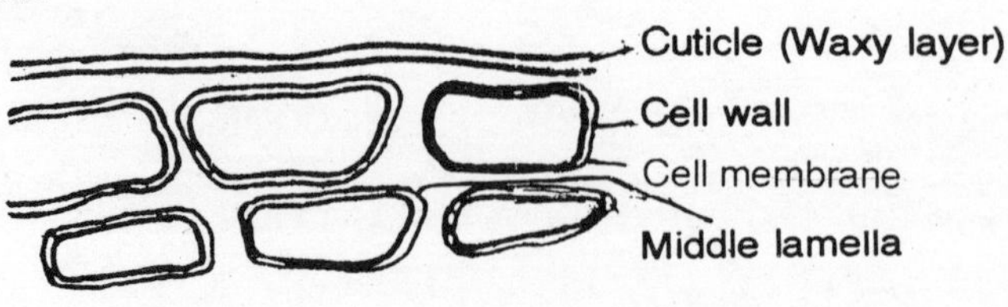

Fig. 46 A plant tissue showing cell wall and middle lamella

Middle lamella is that region where the walls of two cells join in a tissue system. The primary wall is the first wall region formed with definite organization. Secondary wall is that wall portion added after cell elongation is complete.

Cell walls consist of pectic substances, hemicelluloses, cellulose and hydroxyproline rich-proteins. Pectic substances include rhamnogalacturonans, galactans and arabans, hemicelluloses are xylans, xyloglucans, mannans and heteropolymers of glucose, galactose and mannose. Pectic substances are the primary constituents of the middle lamella and the primary cell wall. Hemicelluloses are major constituents of the primary cell wall regions. They link the pectic and cellulosic fractions. Cellulose is found in primary and secondary cell walls, more in the secondary cell walls. In primary wall, cellulose fibrils have a more or less random orientation, but in secondary wall they occur in parallel lamellae. Hydroxyproline-rich proteins found in the cell wall are glycoproteins. The carbohydrate components of this protein are galactose and mannose.

Below the cell wall membranes are found. The membrane provides a semipermeable barrier between the cell and its external environment. The membranes consist of lipids and proteins.

During the biotrophic phase, the pathogens invade the cells by degrading cuticle, middle lamella, cell walls and cell membranes, by producing a series of enzymes. After the biotrophic phase, the pathogens produce toxins which kill the cell and enter into the necrotrophic phase. Thus, enzymes and toxins are important tools of the pathogens for their pathogenesis.

Cutinase

The pathogens produce the enzyme, cutinase which degrades cutin layer. The amount of cutinase released by germinating spores of *Fusarium solani* f. sp. *pisi*, the pea root rot pathogen was correlated with the degree of virulence of various isolates. The avirulent isolate did not produce cutinase. When cutinase enzyme was added to the avirulent isolate, the avirulent isolate also caused disease. When the cuticle was removed from the pea stem tissues, the avirulent isolate even without cutinase enzyme caused the disease (13). All these observations suggest the importance of cutinase in disease development.

Suberin-degrading Enzymes

Suberin-degrading enzymes are produced by pathogens. *Fusarium solani* f. sp. *pisi* is known to produce two suberin-degrading enzymes (10).

Pectic Enzymes

The common pectic substances are of two types:

(i) pectinic acid (pectin), and
(ii) pectic acid (Polygalacturonic acid).

G - COOH	G - CO-O-CH_3
|	|
G - COOH	G - COOH
|	|
G - COOH	G - CO-O-CH_3
|	|
G - COOH	G - COOH
|	|
G - COOH	G - COOH
etc.	etc.
Pectic acid	Pectinic acid
(Polygalacturonic acid)	(Pectin)

The unesterified uronic acid molecules containing up to 75% of methoxyl groups are known as pectinic acid. Those with more than 75% of methoxyl groups are known as pectins. The difference between pectinic acid and pectin is narrow and generally the pectinic acids are called pectins.

Several pectic enzymes are recognized. They are classified based on

(i) the enzyme specificity for a substrate (pectin or pectic acid),

(ii) the mechanism by which the α-1, 4 glycosidic bond is split (i.e. hydrolytic or lyitic), and

(iii) position in the pectic chain at which cleavage occurs (i.e. random or terminal point of attack). The following enzymes are recognized.

1. Pectinmethylesterase (PME)

   It converts pectin into polygalacturonic acid

2. Polygalacturonases

   The enzymes which degrade polyglacturonic acids by hydrolytic action are called polygalacturonases

There are two types of polygalacturonases:

(i) **Exopolygalacturonase:** It cleaves polygalcturonic acids in a terminal manner i.e. releases only monomeric products (galacturonic acids).

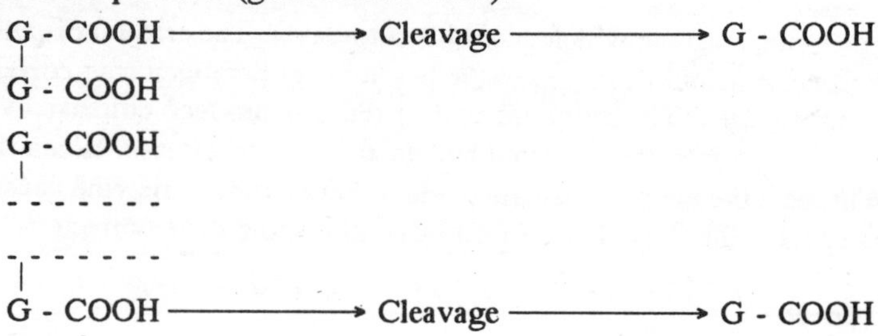

Polygalacturonic acid Galacturonic acid

(ii) **Endopolygalacturonase:** It cleaves polygalacturonic acids in a random manner releasing oligogalou turonic acid.

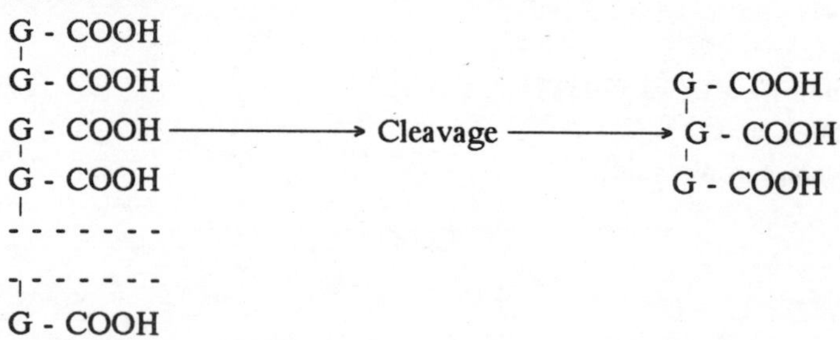

Polygalacturonic acid Oligo-galacturonic acid

(iii) **Pectin methylgalaeturonases:** They act similar to polygalacturonases but the preferred substrate is pectin instead of polygalacturonic acid.

(a) Exo pectin methylgalacturonase - Here the cleavage is terminal

(b) Endo pectin methylgalacturonase - Here the cleavage is at random

(iv) **Polygalacturonate trans-eliminase:** It induces lyitic degradation of the glycosidic linkage resulting in an unstaturated bond between carbons 4 and 5 of a uronide moiety in the reaction product. The transeliminases break the glycosidic linkage at carbon No. 4 and simultaneously eliminating the

H from carbon No. 4, resulting in oligouronides that contain an unsaturated galacturonyl unit (Fig. 47)

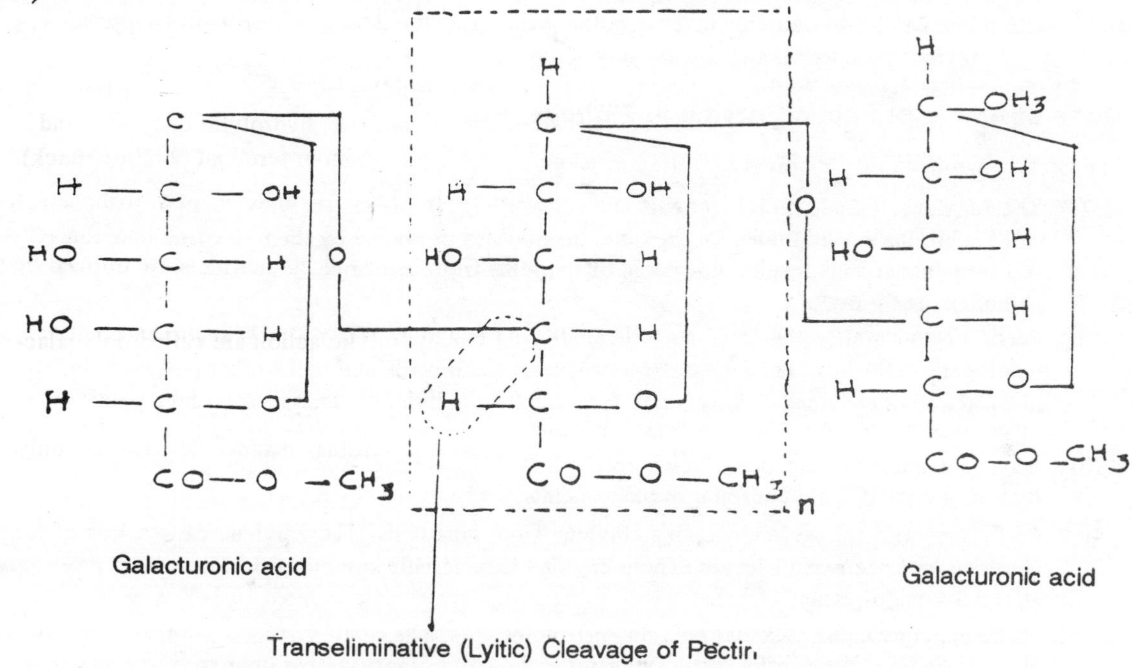

Fig. 47 Transeliminative action of polygalacturonate trans-eliminase

There are two types of polygalacturonate trans-eliminases. They are:

(i) Exopolygalacturonate trans-eliminase - the split is terminal
(ii) Endopolygalacturonate trans-eliminase - the split is at random

These enzymes are also called exo-pectate lyase and endopectatelyase.

(v) **Pectin trans-eliminases:** They are similar to polygalacturonate trans-eliminases but the pectin is the preferred substrate. There are two types of pectin trans-eliminases depending upon 'exo' (terminal) and 'endo' (random) type of reactions.

(i) Exopectin trans-eliminase
(ii) Endopectin trans-eliminase

These enzymes are also called pectin lyase and endo pectin lyase.

Importance of Pectic Enzymes in Pathogenesis

Pathogens are known to produce a variety of pectic enzymes both *in vitro* and *in vivo*. *Fusarium solani* f. sp. *phaseoli,* the bean root rot pathogen produces polygalacturonase, pectin methylgalacturonase, and polygalacturonate trans-eliminase. Pectin methylesterase has been found to be the key enzyme produced by *Fusarium solani* f. sp. *pisi,* the pea root rot pathogen. *Fusarium oxysporum* f. sp. *lycopersici* produces several pectic enzymes, but endo polygalacturonase appears to be more important in pathogenesis. *Verticillium albo-atrun* produces polygalacturonase and pectin trans-eliminase and both of them induce typical symptoms of tomato wilt.

The importance of pectic enzymes has been established by

1. demonstrating that avirulent isolates do not produce or produce less pectic enzymes,

2. by adding the purified enzymes avirulent isolates have been made highly virulent,
3. showing that phenolics and calcium that inhibit pectic enzymes suppress the disease development, and (4) by showing that catabolite (sugars) accumulation in host tissues suppresses pectic enzymes production and disease development.

Mode of Action of Pectic Enzyme in Pathogenesis

The pectic enzymes cause diseases by different ways:

(i) The enzymes degrade middle lamella and cell wall. It facilitates free movement of pathogens from cell to cell. It also facilitates for the toxic metabolites produced by them to come into contact with cell membrane. This results in leakage of nutrients from cells and the nutrients are utilized by the pathogens for growth.

(ii) Pectic enzymes are considered as well-modifying enzymes. They attaçk unaltered plant cell and modify the cell walls. They loosen the structure of cell walls and make other polymers susceptible to the action of enzymes such as cellulolytic and hemicellulolytic.enzymes; without pectic enzymes, other enzymes are not able to attack cell walls.

(iii) Pectic enzymes release peroxidases from cells walls. Peroxidase induces browning of vascular tissues, a typical wilt syndrome in many plants.

(iv) Pectic enzymes release endogenous ethylene from host cells. The ethylene causes loss of turgor, chlorosis and necrosis of leaves. These are the characteristic symptoms in *Verticilium albo-atrum*-affected tomato plants.

(v) Pectic enzymes cause vascular plugging producing gels. The pectic enzymes produced by *Fusarium oxysporum* f. sp. *lycopersici* cause extensive erosion of primary wall of intertracheary pit membrane and vessel secondary wall. These materials result in gel formation. Swelling of the gels leads to occlusion of vessel lumina. It leads to wilting of plants.

(vi) Pectic enzymes in some cases cause cell death. *Erwinia carotovora* produces endo polygalacturonate trans-eliminase and this purified enzyme induces leakage from potato cell membrane and cell death. It suggests that some pectic substances may be present in the cell membrane (19). However; toxic effect of pectic enzymes is uncommon.

(vii) In some cases (*Erwinia carotovora* in soybean and *Rhizopus stolonifer* in castorbean) pectic enzymes release an elicitor from host cells and this elicitor incites phytoalexin synthesis in the hosts (6,7).

Galactanase and Arabanase

The neutral sugar polymers associated with the pectic substances (rhamno-galacturonans) are galactans and arabans. Degradation of galactan is accomplished by exo- and endo-galactanases while araban is degraded by exo-arabanase. These enzymes are produced by pathogens.

Hemicellulases

Many hemicellulosic substances are present in plant cells. They are xylans, xyloglucan, mannan and galactomannan. The important enzymes which degrade hemicelluloses are endoglucanase, β-xylosidase, endoxylanase, endomannanase, β-mannosidase and α-galactosidase. *Sclerotium rolfsii* and *Fusarium* spp. are known to produce these enzymes.

Cellulases

Cellulose molecules consist of long chains of D-glucose residues (1,400 to 10,000 per chain). The linear chains are bound laterally by hydrogen bonds or other physical forces into narrow thread like microfibrils which, in turn, are aggregated to form fibrils. Each microfibril may contain from 280 to 800 cellulose chains. The individual chains in the fibrils vary in degree of orientation. Where the orientation is greatest, the tightly packed parallel chains form crystalline areas. Regions in which chains show a more or less random arrangement are designated amorphous areas. Each linear cellulose chain because of its length passes through several crystalline and amorphous areas (Fig. 48).

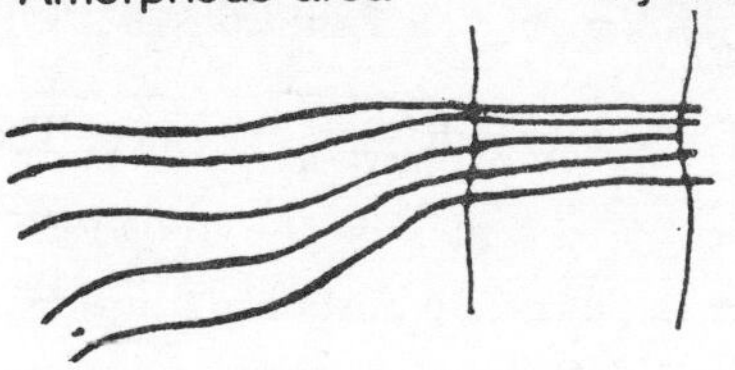

Fig. 48. Structure of cellulose

Cellulose in the cell wall is broken down by a series of enzymes.

(i) **C_1 enzyme:** It acts on native cellulose particularly at crystalline areas which resist the uptake of water because of the tight bonding of the chains. C_1 loosens up the chains so that they take up water prior to their hydrolysis.

(ii) **C_2 enzyme:** It is believed that the enzyme converts cellulose fibres into short insoluble short fibres. Many authors do not recognize C_2 enzyme and they consider it to be only C_1 enzyme.

(iii) **C_x enzymes:** C_x enzymes convert the loosened cellulose fibrils into soluble sugar cellobiose. Many C_x enzymes (C_x A, B, C, etc.) have been identified.

(iv) **β-glucosidase (cellobiase):** It converts cellobiose into glucose.

All these cellulolytic enzymes are produced by plant pathogens. But these enzymes become activated only after the action of pectic enzymes. Cellulases quicky degrade cell walls and facilitate the invasion of host cells by the pathogens.

Hydroxyproline-rich Glycoprotein-degrading Enzyme

Hydroxyproline-rich glycoprotein found in cell walls offers resistance to invasion by the pathogens. Increase in its content in the cell wall confers resistance to many diseases. It has been demonstrated in muskmelon seedlings. The exact role of the enzymes degrading these glycoproteins in disease development has not yet been estblished.

Cell Membrane Degradation

Phospholipases

Phospholipases (phosphatidases) hydrolyze phospholipids present in cell membrane. They degrade phospholipids into fatty acid and a glycerophosphoryl moiety. Four types of phospholipases viz. A, B, C and D are reported to be present. However, the pathogens produce only the first three enzymes while Phospholipase D is produced only by plants.

Phospholipases decompose cell membrane and kill the cells. However, only very few pathogens produce these enzymes.

Proteolytic Enzymes

They hydrolyze peptide bonds in proteins found in cell walls. Many pathogens produce these enzymes. But the role of these enzymes in pathogenesis is not yet studied in detail.

TOXINS

Pathogens produce metabolites toxic to plants. Several authors have tried to classify toxins produced by various pathogens.

Host-specific toxin: It is a metabolic product of a pathogenic microorganism which is toxic only to the host of that pathogen when used at physiological concentrations. At high concentrations even the non-susceptible plants are affected.

Host-selective toxin: Some authors prefer the word 'host-selective toxin' for host-specific toxin because this toxin cannot be strictly called specific as at high concentrations it attacks resistant plants also.

Non-specific toxin: The toxins which are toxic both to hosts and non-hosts of the pathogen are called non-specific toxins.

Pathotoxin: Pathotoxins are the toxins which are major determinants of disease and of pathogenicity. Pathotoxin induces all the typical disease symptoms in reasonable concentration and is correlated with pathogenicity. Host-specific toxin can be considered as pathotoxins.

Vivotoxin: The toxin produced *in vivo* (in the infected tissues) which functions in disease development but not as the initial inciting agent, is termed as vivotoxin.

Phytotoxin: Toxin produced by pathogens, which is toxic to plants but not considered as of primary importance during pathogenesis, is called as phytotoxin. Non-specific toxins are considered as phytotoxins.

Primary determinant: The toxin required for pathogenicity is called primary determinant. It is essential for colonization by the producing pathogen and disease development in the host.

Secondary determinant: The toxin required for virulence is called secondary determinant. Secondary determinant accounts for certain symptoms or contribute to virulence, but are not essential for colonization.

Pathogenicity factor: The primary determinant is otherwise called pathogenicity factor.

Virulence factor: The secondary determinant is otherwise called virulence factor.

Pathogenicity refers to the capacity of an organism to induce disease. Virulence is a quantitative term denoting relative disease inducing ability of a microorganism. A change in host range indicates a change in pathogenicity while a change in severity of disease indicates a change in virulence. But a question arises. When does pathogenesis begin and how much must the pathogen spread for colonization to be successful. Toxin is not required for penetration. Both toxin-producer and non-producer penetrate the host tissues. Initial colonization (upto 24 hr) of both the isolates is also same in case of *H. victoriae,* a powerful toxin producer. Only subsequent invasion is different. It is most likely that toxins play important role only as virulence factor rather than pathogenicity factor. In other words toxin is important in necrotrophic phase rather than in biotrophic phase. All the toxins isolated so far are from saprophytic pathogens. No toxin has been isolated from obligate parasites like rusts, powdery mildews and downy mildews.

Some examples of toxins are given below:

Host-specific Toxins

H.V. toxin: *Helminthosporium victoriae,* the oats victoria blight pathogen produces this toxin. The toxin affects only oat varieties susceptible to the pathogen. Oat varieties with victoria gene alone are susceptible to the pathogen and sensitive to the toxin. Non-pathogenic isolates of the fungus do not produce the toxin. Virulence of the various isolates correlates with the amount of toxin produced in culture. The toxin has been purified. It contains two moieties , one is a peptide and the other moiety is named as victoxinine. Peptide is non-toxic while victoxinine is toxic. But the intact toxin is 7,000 to 10,000 times more toxic than victoxinine. The intact HV toxin is a cyclic polypeptide covalently linked to victoxinine. The molecular weight of the intact HV toxin is about 2000. A mutant of the fungus which has lost the capacity to produce the toxin penetrated the oat leaves and remained confined to the epidermal layer of cells. When HV toxin was added to infection droplets, the mutant then freely colonized the susceptible leaves. The results suggest that the toxin is required to condition susceptible tissues to permit colonization.

It is believed that the toxin is active on susceptible oat leaves because receptor site for the toxin is present only on susceptible tissues. Resistant cells lack the receptor site. The toxin on adsorption by the receptor site acts on cell membrane and induces permeability changes. Electrolytes leak out of the cell.

Helminthosporium maydis-T toxin: Two physiologic races of *H. maydis* cause southern corn blight. The race T is specifically pathogenic to corn containing Texas male-sterile cytoplasm (Cms-T) while race 0 is not. Race T produce a host-specific toxin (pathotoxin) while race 0 produces a non-specific toxin. T toxin has been reported to consist of four components. These components are steroid-like terpenoids and their glycosides. Each component is host-specific and toxic to the same extent. The toxins have been synthesized in the laboratory itself and the synthesized compounds show the same host-specificity.

The toxin causes rapid and drastic changes in mitochondria from susceptible maize, but has no effect on mitochondria from resistant plants. Following toxin treatment, isolated susceptible mitochondria become swollen and respiratory activity of mitochondria is disturbed. Mitochondrial membrane is affected. The toxin prevents open of stomata and inhibits photosynthesis.

Although T-toxin is host-specific, the toxin is not primary determinant of the disease. The fungus penetrates both susceptible and resistant leaves and colonizes both of them. But lesions are larger on susceptible plants. Sporulation on the lesions is also observed on both resistant and susceptible leaves but the sporulation is delayed on resistant leaves. Thus for initial colonization *H. maydis-T* toxin is not essential. T-toxin is only secondary determinant and is a virulent factor. All the highly virulent isolates produce more T-toxin than the less virulent isolates.

Helminthosporium sacchari toxin: *H. sacchari* causes elongated reddish-brown necrotic streaks called 'runner lesions' on sugarcane leaves. Much of the necrotic tissue is free of *H. sacchari* and it is due to the result of the toxin movement, which damages membranes. Host-specific toxin has been isolated from infected tissue. The toxin has been named as helminthosporoside. Its molecular structure is 2-hydroxycyclopropyl α-D-galactopyranoside and its molecular weight is 236. However, the chemical structure of it is in debate for several years. After further purification it has been identified as a mixture of three isomeric bis 5-0 (β-galactofuranosyl)-β-galactofuranosides (A, B and C) differing in the position of one double bond in the sesquiterpene aglycone.

In general, sensitivity or insensitivity to toxin correlated with susceptibility or resistance of the host to the fungus with a few exceptions. The toxin induced electrolyte leakage from cell membrane. A toxin activator has been isolated from susceptible sugarcane leaf washes. It has been identified as serinol. When sugarcane cell cultures are grown with the fungal cultures, toxin production by the fungus is enhanced several fold.

A protein receptor or binding site for helminthosporoside in the cell membrane has been identified. The receptor has been isolated from susceptible sugarcane leaves. The receptor is a protein with a molecular weight of 45,000 to 49,000. A similarly-prepared protein from resistant sugarcane does not bind with toxin.

However, on further purification the receptor from resistant varieties also binds with the toxin. In the presence of digalactosyl diglyceride, a major glycolipid in sugarcane plasmalemma, receptor from resistant varieties loses its binding activity while receptor from susceptible varieties acts even in the presence of the glycolipid.

Three *toxoids* of *H. sacchari* toxin have been isolated from the fungal culture. While the toxin is having four units of galactose per molecule, toxoids I, II and III contain one, two and three units of galactose, respectively. An enzyme with β-galactofuranosidase activity found in the fungal culture converts the toxin to toxoids. Toxoids reduce the activity to toxin. The toxoids are otherwise called *noxins* (14,15).

Toxoids protect sugarcane tissues from toxin. They suppress the electrolyte leakage induced by the toxin. Toxoids occupy the receptor sites competitively with the toxin.

Helminthosporium oryzae-toxin: *H. oryzae* causes brown leaf spot of rice. It produces ophiobolin, ophiobolosin A and ophiobolosin B and all the three are non-specific toxins. They produced symptoms on non-hosts also. However, Vidhyasekaran *et. al.* (1986) isolated a host-specific toxin from the fungus. The toxin elicited the characteristic brown spot symptoms on rice leaves. Rice cultivars susceptible to the pathogens were more sensitive to the toxin than the resistant cultivars. *H. oryzae* could cause infection on sugarcane, sorghum, corn and banana and its toxin also induced typical symptoms on those plants. The toxin does not induce symptoms on non-hosts of the pathogen (30).

When the virulent isolate lost its ability to produce toxin during repeated sub-culturing it became non-pathogenic. When toxin was added to the infection drops of the non-pathogenic isolate, the non-pathogenic isolate also caused infection and colonized rice leaves. The toxin could be isolated from the infected tissues. The toxin has been reported as a primary determinant of the disease (30).

Alternaria mali-toxins: *A. mali* causes blotch of apple. It produces three host-specific toxins. The molecular structures of all the three toxins have been worked out. They are identified as a cyclic depsipeptide (cyclo L - α-amino-p-methoxyphenylvarelyl-dehydroalanyl L-alanyl-L α-hydroxy-isovaleryllactone). These chemicals have also been synthesised in the laboratory. The toxins are released by germinating spores of virulent strains. The toxin induced electrolyte leakage from cell membranes. Susceptible varieties were severely damaged even with 10^{-6} M AM-toxin I while moderately resistant varieties were sensitive only at 10^{-6}M. The resistant varieties were insensitive even to 10^{-4}M toxin.

Two primary sites of action of these toxins have been identified. They attack plasmamembranes and chloroplasts of the susceptible varieties. Chlorophyll content is found to decrease in susceptible varieties treated with AM-toxin I. No such reduction in chlorophyll content has been observed in resistant varieties treated with the toxin.

Other host-specific toxins are:

1. *Helminthosporium carbon* toxin (HC-toxin) — one toxin
2. *Periconia circinata* toxin (PC-toxin) — one toxin
3. *Alternaria alternata* lycopersici toxin (AAL-toxin) — two toxins
4. *Alternaria kikuchiana* toxin (AK-toxin) — three toxins
5. *Alternaria citri* toxin (lemonstrain) (ACR-toxin) — two toxins
6. *Alternaria citri* toxin (tangerine strain) (ACT toxin) — one toxin
7. *Alternaria solani* toxin (AS-toxin) — two toxins
8. *Alternaria alternata* (tobacco strain) toxin (AT-toxin) — one toxin
9. *Alternaria alternata* (Strawberry strain) toxin (AF-toxin) — three toxins

It should be noted that most of the host-specific toxins have been isolated either from *Helminthosporium* or *Alternaria* sp. Both the genera are saprophytic pathogens and whenever they acquire a capacity to produce host-specific toxins, they become highly virulent.

Non-specific Toxins

Fusaric acid: Several *Fusarium* spp. produce fusaric acid in culture. *F. oxysporum* f. sp. *lycopersici,* the tomato wilt pathogen, *F. oxysporum* f. sp. *cubense,* the banana Panama wilt pathogen, *F. oxysporum* f. sp. *vasinfectum,* the cotton wilt pathogen, *F. oxysporum* f. sp. *conglutinans,* the cabbage yellows pathogen, and *F. oxysporum* f. *niveum,* the watermelon wilt pathogen produce fusaric acid in culture as well as in infected tissues. Fusaric acid is considered as 'vivotoxin' as it does not show any host specificity.

Fusaric acid is a pyridine-carboxylic acid (5-butylpicolinic acid). Its molecular weight is 179. Fusaric acid induces epinasty and necrosis of leaves. All the symptoms caused by fusaric acid are not the symptoms of the typical wilt disease caused by the pathogen. Fusaric acid induces wilting of higher plants which are never infected by the fungus. No correlation between the virulence of *Fusarium* isolates and their toxin producing ability has been reported. Although fusaric acid is not responsible for the disease development, it has been reported to be responsible for the intensification of the disease severity. Fusaric acid induces electrolyte leakage of cell membrane. It inhibits glycolate oxidase which is responsible for photosynthesis. Reduction in photosynthesis hastens wilting.

Pyricularia oryzae-toxins (α-picolinic acid, pyricularin, pyriculol and tenuazonic acid): Four non-specific toxins have been isolated from cultures of *Pyricularia oryzae,* the rice blast pathogen. Pyricularin, α-picolonic acid and tenuazonic acid have been isolated from the diseased rice leaves also. All the four toxins, pyricularin, α-picolinic acid, pyriculol and tenuazonic acid individually induced typical blast lesions on rice leaves (11, 29). Pyricularin and α-picolinic acid have been detected even in germinating conidia. Varietal reaction to the fungus is not correlated with resistance to these toxins. Treatment of rice plants with pyricularin (10 μg/ml) reduced the resistance of the host cell to hyphal invasion and upsets the functioning of the cell.

Phaseotoxin and Phaseolotoxin

Pseudomonas syringae pv. *phaseolicola* causes halo blight disease in beans. The bacterium produces a toxin called as phaseotoxin and the toxin induces characteristic symptoms of the disease, viz. chlorotic halos in the leaves. The toxin is a peptide yielding 5 amino acids upon acid hydrolysis. In infected tissues ornithine accumulates with inhibition of citrulline and arginine. Inhibition of this ornithine cycle results in chlorosis. The toxin inhibits the enzyme ornithine carbamoyl transferase (OCT) which results in inhibition of synthesis of citrulline. The toxin-induced chlorosis is reversed by application of citrulline or arginine.

The bacteria multiply substantially even in resistant varieties without production of phaseotoxin. Large quantities of phaseotoxin are detected in susceptible varieties. When the resistant varieties are treated with phaseotoxin, the multiplication of the bacteria increases several fold and typical disease symptoms appear. It suggests that the resistant varieties have some inhibitors which inhibit the production of phaseotoxin.

Phaseotoxin suppresses defensive mechanisms of bean plants. It suppresses synthesis of phytoalexins (phaseollin, phaseollinisoflavan, and kievitone) which are inhibitory to the bacteria. Thus, the primary role of the toxin appears to suppress the host defense mechanisms. Once the toxin accumulates, disease symptoms are produced. This type of toxins can be called as 'aggressins'. Aggressins are the chemicals which breach host defenses (26).

Another toxin, phaseolotoxin has also been isolated from cultures of *P. syringae* pv. *phaseoli.* This toxin has much more potential to induce chlorosis than phaseotoxin. Phaseolotoxin is also a peptide containing 3 amino acids which are entirely different from those of phaseotoxin. The structure of phaseolotoxin is (n^3-phosphosulphamyl) ornithylalanylhomoarginine. Phaseolotoxin has been observed to be the major toxin and more than 80 per cent of toxin activity in culture medium has been ascribed to this toxin. Phaseolotoxin also induces ornithine accumulation and induces typical halo blight symptoms.

A third toxin has also been isolated and it is an analogue of phaseolotoxin. It is called (2-serine) - phaseolotoxin. It also induces typical symptoms of the disease.

Tabtoxin and 2 serine-tabtoxin

Pseudomonas syringae pv. *tabaci,* the causal agent of wild fire of tobacco produces two toxins, viz. tabtoxin and 2-serine-tabtoxin. They are dipeptides. Tabtoxin is α-lactamthreonine while 2-serine tabtoxin is α-lactamserine. Their molecular weights are 289 and 275 respectively. Both the toxins are collectively called as tabtoxins. The wildfire disease is characterized by loss of chlorophyll, accumulation of ammonia and inactivation of glutamine synthetase. Tabtoxin induces chlorotic haloes in tobacco leaves. Earlier studies suggested that

(i) tabtoxin acts as an antagonist of L-methionine similar to methionine sulfoximine, an analogue of L-methionine, and

(ii) tabtoxin inhibits glutamine synthetase. The first theory is not now accepted. The second theory is being questioned. The third theory is that tabtoxin induces ammonia accumulation in the leaves. Ammonia regulates the Kreb's cycle and it may induce necrotic symptoms which occur before the chlorosis formation. Inhibition of glutamine synthetase may lead to accumulation of ammonia.

Even toxinless mutants of *P. syringae* pv. *tabaci* invade tobacco cells and multiply well. But they don't produce symptoms of the disease. The toxinless mutants can be called 'non-pathogenic parasites'. The studies suggest that tabtoxin is a major determinant of the pathogenicity of *P. syringae* pv. *tabaci.* Without the toxin, the bacteria do not produce disease symptoms (28).

Tabtoxin is produced by other bacteria also. *Pseudomonas coronafaciens,* the oat bacterial blight pathogen and *Pseudomonas* sp, the pathogen of timoty also produce tabtoxin. Tabtoxin is a non-specific toxin.

Other non-specific Toxins

Several other pathogens produce non-specific toxins.

Cercosporin is produced by several *Cercospora* spp. Tenuazonic acid is produced by several *Alternaria* spp. Collectotion is produced by *Colletotrichum* spp. Ophiobolin is produced by several *Helminthosporium* spp.

Extracellular polysaccharides (EPS) are produced by several pathovars of *Xanthomonas campestris* (pv. *oryzae,* pv. *phaseoli* etc.), *Verticillium* spp., *Corynebacterium* spp. and *Erwinia amylovora.*

Although these toxins are all non-specific, they also play an important role in pathogenesis.

ENZYMES WHICH DEGRADE HOST DEFENSE CHEMICALS

Many toxins are known as aggressins. They breach the host defense barrier. They suppress the synthesis of defense chemicals. Pathogens produce another set of enzymes which degrade or detoxify the defense chemicals of the plants. Thus toxins suppress the synthesis of defense chemicals while enzymes detoxify those defense chemicals.

Tomatine Hydrolase

α-Tomatine is a glycoalkaloid found in tomato plants. It is inhibitory to fungi. However, the pathogen *Fusarium oxysporum* f. sp. *lycopersici* produces an enzyme which hydrolyses tomatine into a less toxic substance, tomatidine. Tomatidine is not inhibitory to the pathogen.

Pisatin Demethylase

Pisatin is the important phytoalexin in pea. *Fusaium solani* f. sp. *pisi,* the pea root rot pathogen is not inhibited by pisatin. All the highly virulent isolates produce pisatin demethylase and degrade pisatin to 3, 6a dihydroxy-8, 9-methylene dioxypterocarpan which is non-fungitoxic. Less virulent isolates lack the ability to produce the enzyme. Pisatin demethylase has been identified has cytochrome monooxygenase (16).

Phaseollin Monooxygenase

Phaseollin is an important phytoalexin in bean. *Fusarium solani* f. sp. *phaseoli,* the bean root rot pathogen degrades phaseollin to a less toxic product, 1a-hydroxyphaseollone by producing an enzyme, phaseollin mono-oxygenase (12).

Klevitone Hydratase

F. solani f. sp. *phaseoli* degrades kievitone, another phytoalexin of beans to kievitone hydrate which is less toxic. It produces the enzyme kievitone hydratase. All the virulent isolates of the pathogen produce this enzyme while non-pathogens do not produce (with a few exceptions). This enzyme is detected in the pathogen-inoculated bean tissues also (25).

Several virulent pathogens have been reported to produce enzymes which degrade phenolics and phytoalexins of host tissues. Thus a successful pathogen produces an array of enzymes and toxins, overcomes the defense barrier of plants and causes severe disease. Similarly host produces an array of compounds which inhibit the enzymes of the pathogen and detoxify the toxins produced by them. The battle between pathogen and host is everlasting. If the pathogen wins, the disease occurs and if the host wins, the disease is suppressed. Man should learn to help the host winning this battle. Disease management by the human effort is given in subsequent chapters.

REFERENCES

1. Bateman, D.F. and H.G. Basham, 1976. Degradation of plant cell walls and membranes by microbial enzymes. In R. Heitefuss and P. H. Williams. Encyclopaedia of Plant Physiology New Series, Physiological Plant Pathology Vol. **4**: 316-355, Springer-Verlag, New York
2. Bruce, R. J. and C. A. West. 1982. Elicitation of casbene synthetase activity in castor bean. The role of pectic fragments of the plant cell wall in elicitation by a fungal endopolygalacturonase. Plant Physiol, **69**: 1181-1188
3. Collmer, A. and N. T. Keen, 1986. The role of pectic enzymes in plant pathogenesis. Annu. Rev. Phytopathol, **24**: 383-409
4. Daly, J.M, 1982. The host-specific toxins of Helminthosporia. In: Y. Asada, W. R. Bushnell, S. Ouchi and C.P. Vance (ed.) Plant infection: The physiological and biochemical basis. Japan Sci. Soc. Press, Tokyo/ Springer-Verlag, Berlin, pp. 215-234
5. Daly, J.M. and I. Uritani., 1979. Recognition and specificity in plant host-parasite interactions. Japan Scientific Press Tokyo and University Park Press, Balitmore, p. 355
6. Davis, K.R., G.D. Lyon, A.G. Darvill and P. Albersheim, 1984. Host-pathogen interactions. XXV Endo-polygalacturonic acid lyase from *Erwinia carotovora* elicits phytoalexin accumulation by releasing plant cell wall fragments, Plant Physiol, **74**: 52-60
7. Defago, G. and H. Kern, 1983. Induction of *Fusarium solani* mutants insensitive to tomatine, their pathogenicity and aggressiveness to tomato fruits and pea plants. Physiol Plant Pathol, **22**: 29-37

8. Durbin, R.D. 1981. Toxins in plant disease. Academic Press, New York, p. 515
9. Duvick, J.P., J.M. Daly, Z. Kratky, V. Macko, W. Acklin and D. Arigoni, 1984. Biological activity of the isomeric forms of *Helminthosporium sacchari* toxin and of homologs produced in culture. Plant Physiol, **74**: 117-122
10. Fernando, G., W. Zimmermann and P. Kolattukudy, 1984. Suberin-grown *Fusarium solani* f. sp. *pisi* generates a cutinase-like esterase which depolymerizes the aliphatic compounds of suberin. Physiol. Plant Pathol, **24**: 143-155
11. Iwasaki, S., R. Nozoe, S. Okuda, Z. Sato and T. Kozaka, 1969. Isolation and structural elucidation of a phytotoxic substance produced by *Pyricularia* oryzae cavara. Tetrahedron Letters, **45**: 3977-3980
12. Kistler, H.C. and H.D. VanEtten, 1981. Phaseollin metabolism and tolerance in *Fusarium solani* f. sp. *phaseoli*, Physiol. Plant Pathol, **19**: 157-271
13. Koller, W., C. R. Allan and P.E. Kolattukudy, 1982. Role of cutinase and cell wall degrading enzymes in infection of *Pisum satium* by *Fusarium solani* f. sp. *pisi.*, Physiol. Plant Pathol, **20**: 47-60
14. Livingston, R.S. and R.P. Sheffer, 1981. Isolation and characterization of host-selective toxin from *Helminthosporium sacchari*, J. Biol. Chem., **256**: 1705-1710.
15. Livingston, R.S. and R.P. Scheffer, 1984. Toxic and protective *effects of analogues of Helminthosporium sacchari* toxin on sugarcane tissues. Physiol Plant Pathol, **24** : 133-142
16. Matthews, D.E. and H.D. Van Etten, 1983. Detoxification of the phytoalexin pisatin by a fungal cytochrome p-450. Arch. Biochem., Biophysics, **224**: 494-505
17. Mirocha, C.J. and I. Uritani, 1967. Dynamic role of molecular constituents in plants-parasite interaction. Bruce Publ. Co. St. Paul, p. 372
18. Mitchell, R.E., J.S. Johnston and A.R. Ferguson, 1981. Phaseolotoxin and other phosphosulphamyl compounds: Biological effects. Physiol. Plant Pathol, **19**: 227-235
19. Mount, M.S., D.F. Bateman and H. Grant Bascham, 1970. Induction of electrolyte loss, tissue maceration and cellular death of potato tissue by an endopolygalacturonate trans-eliminase. Phytopathology, **60**: 924-931
20. Nishimura, S. and K. Kohmoto, 1983. Host-specific toxins and chemical structures from *Alternaria* species. Annu. Rev. Phytopathol, **21**: 87-116
21. Patil, S.S. 1974. Toxin produced by phytopathogenic bacteria. Annu. Rev. Phytopathol, **12**: 259-279
22. Pringle, R.B. and R.P. Scheffer, 1964. Host-specific plant toxins. Annu. Rev. Phytopathol, **2**: 133-156
23. Rudolph, K. 1976. Non-specific toxins, In: R. Heitfuss, and P. P. H. Williams (ed.) Physiological Plant Pathology, Encyclopaedia of Plant Pathology, Springer-Verlag, Heidelberg, New York, pp. 270-315
24. Scheffer, R.P. and S.P. Briggs. 1981. Introduction: A perspective of toxin studies in plant pathology. In: R.D. Durbin (ed.). Toxins in Plant Disease. Academic Press, New York, p. 1-20
25. Smith, D.A., J.M. Harrer and T.E. Cleveland, 1982. Relation between production of extracellular kievitone hydratase by isolates of *Fusarium* and their pathogenicity on *Phaseolus vulgaris.* Phytopathology, **72**: 1319-1323
26. Smith, H, 1968. Biochemical challenge of microbial pathogenicity. Bacterial Rev. **32**: 164-184
27. Strobel, G.A., 1982. Phytotoxins. Annu. Rev. Biochem, **51**: 309-333
28. Turner, J.G. and R. R. Taha, 1984. Contribution of tabtoxin to the pathogenicity of *Pseudomonas syringae* pv. *tabaci.* Physiol. Plant Pathol, **25**: 55-69
29. Umetsu, N., J. Kaji and K. Tamari 1972. Investigation on the toxin production by several blast fungus strains and isolation of tenuazonic acid as a novel toxin. Agrl. Biol. Chem., **36**: 859-866
30. Vidhyasekaran, P., E.S. Borromeo and T.W. Mew, 1986. Host-specific toxin production by *Heminthosporium oryzae.* Phytopathology, **76**: 261-266

31. Wood, R.K.S., A. Ballio and A. Graniti, 1972. Phytotoxins in plant diseases. Academic press, London and New York, p. 530

32. Wheeler, H. and H.H. Luke, 1963. Microbial toxins in plant diseases. Annu. Rev. Microbiol **17**: 223-242

33. Yorder O.C., 1980. Toxins in pathogenesis. Annu. Rev. Phytopathol, **18**: 103-129.

12

PHYSIOLOGY OF DISEASE RESISTANCE

Disease is rather exception than a rule in plants. Plants have defense mechanism against almost all fungi, bacteria and viruses and as exception a few pathogens circumvent these defense mechanisms and cause diseases. The defense mechanisms of host plants are presented in this chapter.

Recognition

The host recognizes pathogens and non-pathogens within a short period (15 minutes to few hours). For example, pea plant recognizes an incompatible pathogen *Fusarium solani* f. sp. *phaseoli* (a pathogen of bean) within 3 hrs. of inoculation and suppresses growth of the non-pathogen. On recognition the plant:

(i) retards spore germination of the non-pathogen, or
(ii) reduces appressorium formation, or
(iii) prevents penetration of infection hypha, or
(iv) retards growth rate of infection hypha inside the host tissue, or
(v) inhibits haustoria formation, depending upon the host-plant interaction. In case of bacteria, the plant forms a bacteria-limiting wall cuticle (an envelope) around the bacteria and immobilizes the bacteria within about 6 hrs. (28, 29).

Messenger Ribonucleic Acid (mRNA)

For any action, a message from deoxyribonucleic acid (DNA) should arise as DNA is the basic heritable unit of life. The message from DNA is carried by mRNA. The message carried by mRNA is transcribed by transfer RNA (tRNA) and translated by ribosome RNA (rRNA). When an incompatible pathogen invades the host, synthesis of all the RNAs increases and particularly mRNA synthesis increases several folds (5). The increase in mRNA synthesis is observed even within 15-30 minutes of the fungal invasion. Normally, it takes more than an hour for the fungal spores to germinate. It suggests that a mere contact between fungal spore and host cell is sufficient to initiate the defense mechanisms.

Protein Synthesis

Immediately followed by an increase in mRNA synthesis, protein synthesis increases in the host on invasion by an incompatible fungus. The enzymes are proteins. The increased synthesis of proteins suggests the activation of enzymes which are essential for the synthesis of various defense chemicals.

Phenolics

Phenolics are the common compounds found in plants and possess an aromatic ring bearing a hydroxyl substituent. Phenolics are toxic to fungi, bacteria and viruses. They inhibit:

(i) spore germination,
(ii) mycelial growth,
(iii) production of pectic and cellulolytic enzymes,
(iv) activity of pectic and cellulolytic enzymes, and
(v) production of toxins by the fungi. They also detoxify the phytotoxins produced by pathogens.

Normally the phenolic contents of the plants are not very high, sufficient enough to suppress the pathogens. Phenolics are toxic to plant cells also besides the pathogens. Hence phenolics do not accumulate in healthy uninfected plants. But when an incompatible pathogen invades the host, on recognition [probably by activation of mRNA and protein (enzymes)] synthesis of phenolics is activated and phenolics accumulate at the site of infection. This increase in phenolics content has been found to be due to increase in synthesis of phenylalanine ammonialyase (PAL), a key enzyme in the metabolism of phenolics. Some of the individual phenolic substances have been identified as the key defense chemicals in plants. They are:

1. Catechol and pyrocatechuic acid in onion,
2. Chlorogenic acid in potato,
3. Umbelliferone and scopoletin in sweet potato,
4. Catechin, gallocatechin and isoquercitin in cotton,
5. Avenacin in oats,
6. Phloridzin in apple,
7. Hordatine in barley,
8. Tomatine in tomato,
9. Methoxymellein and falcarinol in carrot,
10. Tuliposides in tulip.

Quinones

Phenolics are oxidized to quinones by the host enzymes, polyphenol oxidase and peroxidase. Quinones are much more toxic than phenolics. Quinones have been reported to be responsible for disease resistance in many crops. Quinones are known to inactivate viruses. They combine with virus proteins and 'tanning of viruses' leads to suppression of virus multiplication.

Phytoalexins

Plants synthesize some new defense chemicals when they are invaded by micro-organisms. They do not exist in plants prior to microbial invasion. The newly synthesized chemicals are called phytoalexins.

Phytoalexins are defined as low molecular weight, antimicrobial compounds that are synthesized by and accumulated in plants after exposure to microorganisms. Phytoalexins are also defined as antibiotics formed in plants via metabolic sequence induced either biotically or in response to chemical or environmental factors (18).

Phytoalexins are all phenolics in nature. They are mostly isoflavonoids, terpenoids, and polyacetylene compounds. Phenylalanine ammonia lyase (PAL) is the enzyme involved in the biosynthesis of phytoalexins. It is believed that dying plant cells (necrosis) may release a diffusive factor which acts as a trigger of phytoalexin synthesis in surrounding cells. In short, phytoalexins are produced when the plants are in stress conditions and they can be called 'stress-compounds'.

Phytoalexins inhibit:

(i) spore germination,
(ii) germ tube elongation,
(iii) growth of sporeling (young mycelia from germinating spores), and
(iv) mycelial growth. The phytoalexins are toxic mostly to terminal cells of the hyphae. They are inhibitory to bacterial growth and are toxic even to plant cells.

The role of phytoalexins in disease resistance has been studied in detail (10). Their role varies from plant to plant and even among each plant-pathogen interaction. The following conditions have been reported in various plant-pathogen interactions:

(i) Phytoalexins are produced in resistant interactions only,
(ii) Phytoalexins are produced in both resistant and susceptible interactions but more in the resistant ones,
(iii) Phytoalexins are produced more rapidly in resistant reactions than in the susceptible ones,
(iv) Incompatible pathogens induce more phytoalexins than the compatible ones,
(v) Phytoalexins are degraded by pathogens and not by incompatible ones,
(vi) Phytoalexins are degraded by both pathogens and non-pathogens but rapidly by pathogens,
(vii) Phytoalexins are degraded by the susceptible host itself,
(viii) Non-pathogens are more sensitive to the phytoalexins than the pathogens,
(ix) Pathogens adapt to the phytoalexins quickly compared to non-pathogens,
(x) Phytoalexins are produced at the appropriate sites at appropriate concentrations and time in the resistant reactions.

All the ten possibilities may exist in the one and the same host-pathogen interactions or one or two alone may be responsible. Existence of any one of the factors can confer resistance. But there may be cases in which all the ten possibilites may not exist and the phytoalexins may not be responsible for disease resistance at all. They may be merely stress compounds.

Many phytoalexins have been isolated and characterized from host plants. Some of them are:

1. Pisatin in pea,
2. Phaseollin, phaseollidin, phaseollinisoflavan, coumestrol, and Kievitone in bean,
3. Momilactone A and B, and oryzalexins A, B and C in rice,
4. Rishitin and 5 unidentified terpenoids in tomato,
5. Rishitin and phytuberin in potato,
6. Capsidiol and capsicannol in chilli,
7. Hemigossypol (isohemigossypol), methoxy-hemigossypol, vergosin (desoxymethoxy hemigossypol) and gossyvertin in cotton,
8. Glyceollin in soybean,
9. Medicarpin and maackiain in chickpea,
10. Ipomeamarone, dehydroipomeamarone and ipromeamarone in sweet potato,
11. Avenalumin I and II in oats,
12. Wyerone acid in broad bean,
13. ξ- viniferin, ξ-viniferin and pterostilbene in grapevine,
14. Sativan and medicarpin in lucerne.

Elicitors of Phytoalexins

A fungal or bacterial genome interacts with host genome and mediates the activation of a biosynthetic pathway previously quiescent (dormant) in the healthy host which results in the synthesis of phytoalexins. Some specific compounds have been isolated from various fungi and bacteria which induce the synthesis of phytoalexins (2). These molecules of fungal and bacterial origin which initiate the accumulation of phytoalexins in plant tissues are called 'elicitors'.

The elicitors are found in outer mycelial cell walls. They are mostly glycoproteins. Elicitor isolated from *Cladosporium fulvum,* the tomato pathogen consists of mannose, galactose and protein. A glucan has been identified as elicitor of *Phytophthora infestans,* the potato late blight pathogen. The elicitor activates messenger RNA which induces synthesis of phenylalanine ammonia lyase (PAL).

Suppressor

The fungi which produce elicitors, also produce suppressors which suppress the phytoalexin synthesis in hosts. The suppressors like elicitors are also glycoproteins. The suppressors of *Phytophthora infestans* is also a glucan similar to the elicitor. However elicit or of *P. infestans* is a high molecular weight glucan while suppressor from the fungus is a low molecular weight glucan.

Receptor

Both elicitors and suppressors of phytoalexins are present in the cell wall of pathogens. For their action, there should be receptors in the host. Elicitors may bind with receptor on plant cell membranes and elicit phytoalexin synthesis. Suppressor may bind with host receptors which in turn would prevent the availability of receptor site for the action of elicitor.

It is postulated that in case of resistant reactions the elicitor of incompatible pathogens binds to the receptor. This results in synthesis of a specific mRNA which activates appropriate regulatory loci that trigger the expression of many structural genes associated with resistance. In case of susceptibility, the suppressors of compatible pathogens may occupy the receptor sites and make them unavailable to the elicitors.

Lignins

Lignins are the most abundant and widely distributed phenolic polymers found in plants. Quantitatively lignins far surpass not only other plant phenolics but also all other organic natural products except cellulose. Lignin plays an important role in disease resistance (21).

Lignin may act as a physical barrier for the entry of incompatible pathogens into the host. When the fungus penetrates, a tubular structure develops around the penetration hyphae and they are called 'papillae'. The papillae are made up of lignin. Besides the papillae, the upper epidermal cell walls are also lignified. These lignified cells prevent the entry of the incompatible pathogens into the host. This type of lignified barrier results in thickening of cell wall and it prevents cell to cell spread of viruses. In some cases the invading hyphae themselves will be lignified and the lignified hyphae cannot make further growth.

Callose

Callose is the substance found in sieve tubes and prevents the leakage of sieve tube sap or water in the cell walls. It is chemically related to cellulose and proteinaceous in nature. Callose appears rapidly in the cell walls of parenchyma cells when these are wounded

Similar to lignin, callose also prevents entry of incompatible pathogens into the host. Papillae formed in the outer epidermal walls beneath penetration pegs consist of callose besides lignin. Callose restricts the passage of small molecules between tissues. Hence deposition of callose completely seals the invaded cells and prevents the movement of nutrition from cells to pathogens and movement of toxins from pathogens to the host cells.

Callose deposits are found in plasmadesmata when incompatible virus is inoculated. These deposits seal off the plasmodesmata. Viruses move from cell to cell only through plasmodesmata. Thus sealing off the plasmodesmata by callose prevents virus spread inside the host.

Lytic Enzymes

Chitins and β- 1, 3 glucans are the major components of cell walls of most of the fungi. Chitinase and glucanase are the two enzymes which degrade chitins and glucans of the fungi. These enzymes are called lytic enzymes because they cause the lysis of fungi. These enzymes are produced by both plants and fungi. Autolysis of fungal mycelia by the enzymes produced by themselves is common particularly in old cultures. But in the host-pathogen interactions host lytic enzymes may play an important role. In resistant varieties and incompatible interactions, the plant produces more lytic enzymes and the fungal mycelium in the host tissue is lysed.

Chitinase acting on the cell wall of fungi releases chitosan which is a derivative of chitin. This chitosan induces phytoalexin synthesis in the plants. Chitosan itself is toxic to fungi (7). Thus chitinase of the host inhibits growth of fungi by:

(i) activating phytoalexin synthesis,
(ii) by the inhibitory action of chitosan, and
(iii) by lysing the mycelium.

Peroxidase

It is an important enzyme found in plants. In many plant systems peroxidase has been correlated with disease resistance. Peroxidase is an important enzyme in the synthesis of lignins. It also catalyzes the oxidation of phenols into quinones which are highly fúngitoxic. Peroxidase itself is toxic to some pathogens. It has antibacterial activity

Polyphenoloxidase

Polyphenoloxidase is another enzyme commonly found in plants. It oxidizes phenols to highly fungitoxic quinones. The enzyme itself is inhibitory to some plant viruses. It inactivates RNA of the virus.

Glycosidase

Phenolics are not only fungitoxic but also phytotoxic. High accumulation of phenolics in plants will retard the growth of plants. But phenolics are essential for their defense reaction. Hence phenolics are found stored in the form of glycosides (phenolics + glucose or galactose) which are less toxic. At the time of infection, host glycosidases may convert the glycosides into the toxic aglycones (phenolics) and prevent the spread of the pathogen. In this way host glycosidases may play an important role in disease resistance.

Superoxide anion (O_2^-)

NADPH oxidase is the enzyme taking part in terminal respiration (oxidative phosphorylation) pathway in plants. Its activity generates a superoxide anion (O_2^{-}) in potato tissues. This superoxide anion has been shown to induce phytoalexins in potato. Superoxide anion accumulates more in incompatible (resistant) interaction. This accumulation leads to accumulation of more phytoalexins.

Auxins

Auxins are plant growth regulators. Different cells in plant tissue are bound tightly by pectin. Auxins alter the nature of pectin and make them either water-soluble or insoluble. Water-soluble pectins are easily attacked by pathogens. In resistant plants, auxins accumulate and make the pectins water-insoluble. Auxins can increase phenolics synthesis and induce resistance. Auxins can also modify sugar content of the plants and sugars also can induce resistance (see under 'sugars'). Indole Acetic Acid (IAA) is the important auxin found in plants.

Ethylene

In some resistant host-pathogen interactions, ethylene is released abundantly from invaded tissues. Ethylene induces resistance in different ways;

(i) Ethylene increases peroxidase and polyphenol oxidase activities in plants. We have already seen the role of peroxidase and polyphenoloxidase in disease resistance.
(ii) Ethylene increases synthesis of phenolics which are inhibitory to fungi, bacteria and viruses,
(iii) Ethylene induces phytoalexin synthesis.

Ascorbic acid

Ascorbic acid content in plant varies from plant to plant, but it is usually high. This acts as the substrate for ascorbic acid oxidase, one of the enzymes which are predominant in cyanide-resistant respiration. The respiration of normal plants is cyanide-sensitive because cytochrome oxidase is rapidly inactivated by reaction with cyanide (1).

Ascorbic acid is needed for the development of cyanide-resistant respiration (17). The resistance is correlated with the development of cyanide-resistant respiration which develops earlier and to a larger extent in the resistant cultivars.

Sugars

The diseases have been classified as 'low sugar diseases' and 'high sugar diseases'. Low sugar diseases are those which occur severely when the host sugar content is low and high sugar diseases are those which occur severely when the host sugar content is high. Leaf spots caused by *Helminthosporium* spp. and *Alternaria spp.* are considered as low sugar diseases while rusts and powdery mildews are considered as high sugar diseases. The mode of action of sugars in inducing resistance may be as follows:

(i) Sugars are the important carbon sources for nutrition of pathogens. Hence increase or decrease in sugar content of plant tissues may affect the disease development.
(ii) Sugar concentration in the tissues is a key factor in toxin production by the pathogens.
(iii) Pectolytic and cellulolytic enzymes of pathogen origin help the pathogens to penetrate into the cells or to develop in middle lamella region (intra- and inter-cellular growth). Pectin and cellulose, the substrates for these enzymes are complex polysaccharides. The pathogens prefer simple sugars rather than polysaccharides. Hence if sugars accumulate more in plant cell walls, the production of these enzymes will be suppressed due to preferrential utilization of sugars. If a plant presents to a pathogen, a mixture of sugars which leads to the insufficient secretion

or wrong timing of secretion of polysaccharide-degrading enzymes by the pathogen, abortive development of the pathogen may result. This would render the plant resistant to infection by the pathogen.

(iv) Sugars are the precursors of phenolics. Higher sugar content results in higher synthesis of phenols (23).

Amino Acids

Numerous amino acids occur freely in abundance in plants. Pathogens prefer amino acids for their growth. The amino acids may induce resistance by the following ways:

(i) Phenylalanine and tyrosine are the two phenolic amino acids which are precursors for the synthesis of phenolics.

(ii) Methionine, methyl aspartic acid and leucine are amino acids which contain a terminal methyl group. These methyl groups may alter the nature of pectin in middle lamella and induce resistance.

Hydrocyanic Acid

Some of the plant species like sorghum and flax contain more hydrocyanic acid. Those plants are called cyanogenic plants. Hydrocyanic acid (HCN) is released only when the host cells are damaged. When the pathogen invades and kills the cell, hydrocyanic acid is released and it kills the pathogen. HCN is found responsible to induce resistance against certain diseases.

Calcium

Calcium combines with pectic acid and forms calcium polypectate. Calcium polypectate is resistant to pectic enzymes of the pathogen. The pathogens generally alter host cell membrane permeability. Leakage of nutrients through cell membrane permits growth of pathogens in the inter-cellular region. Prevention of electrolyte leakage will result in resistance. Calcium retards deterioration of cell membrane and suppresses electrolyte leakage. Thus calcium in plant tissues induces resistance:

(i) by altering pectin metabolism, and

(ii) by increasing cell membrane turgidity.

Iron

Iron is an essential co-factor for the enzymes activating synthesis of phytoalexins. Iron accumulation increases phytoalexin synthesis in chilli. Iron decreases virulence of pathogens in some cases. Thus, iron may induce resistance:

(i) by increasing phytoalexin synthesis in plants, and

(ii) by decreasing virulence of the pathogen.

Copper

Copper is a co-factor for peroxidase. Increased copper content may activate peroxidase and induce resistance.

Silicon

Silicon accumulates in the resistant interaction in some hostpathogen systems. Silicon suppresses haustorium formation by the pathogen.

Anti-Viral Factor (AVF)

Resistance to virus diseases is observed when an incompatible virus infects the host. Several chemicals like polyacrylic acid, aspirin, salicylic acid, benzoic acid and various growth regulators like Indole acetic acid (IAA), 2, 4 -D, Benzyl amino-urine (BAP) and ethylene induce resistance against virus diseases. Various plant products also induce resistance. In all these cases new proteins have been found to develop in plant tissues. They are called 'antiviral factor'. These antiviral factors do not exist in healthy plants. They are produced only when the virus infects the host. They are post-infectional compounds similar to 'phytoalexins'. In tobacco several antiviral factors have been detected. As many as ten new proteins have been detected in tobacco mosaic virus inoculated tobacco cultivars. These antiviral factors have been termed as 'Resistance Associated Proteins' (RAPs) or 'Pathogenesis Related Proteins' (PRP) by different authors.

In cucumber, resistance against cucumber mosaic virus observed in resistant varieties has been related to the synthesis of antiviral factors activated by virus infection. An antiviral factor has been isolated also from tomato plants resistant to tobacco mosaic virus after infection with the virus.

The anti-viral factor does not inhibit the infectivity of viruses. It also does not affect the process of 'uncoating of coat protein'. The AVF inhibits only the multiplication and spread of viruses.

Anti-Viral Principle (AVP)

Anti-viral factor is extracted from virus-infected tissues and not from resistant uninfected plants. But in a few instances, some antiviral substances have been isolated from the systemically induced resistant leaves without challenge inoculating with viruses. This substance is called 'Anti-viral Principle' (AVP). Increased inhibitory activity is found in saps from the resistant bean leaves when other parts of the plants are inoculated with tobacco mosaic virus. The inhibitory activity has been found only in the purified RNA preparation. It suggests that the antiviral principle may be a RNA.

Lectins

Lectins are either proteins or glycoproteins which bind specific saccharides. Lectins are present in membranes of plant cells and cell walls. Since, they are present on the outer cell surfaces of the host, they may act as recognition sites for pathogens or non-pathogens. Many incompatible pathogens have been reported to be bound with lectins of the host while pathogens remain unattached to the lectins of host cell. Lectins may act as receptors for elicitors of phytoalexins present in the surface of cells of the incompatible pathogens. When the elicitors bind with the receptors (lectins), disease resistance is induced. Lectins may inhibit chitin formation in hyphal tips resulting in inhibition of growth of incompatible pathogens.

The role of lectins in early recognition process has been established only in a few host-pathogen interactions. In some others lectins have been found to have no role in recognition process (20).

Hypersensitivity

In many incompatible host-pathogen interactions host cells are rapidly killed around the sites of penetration. The rapid necrosis observed in the incompatible interactions is termed as Hypersensitive Response (HR) or hypersensitivity. It is believed that hypersensitive cell death induces resistance. Hypersensitive response has been observed even within 2 hrs. after inoculation with the pathogen in the resistant reaction. Thus host cells die long before the cessation of hyphal growth in resistant varieties suggesting that cell death may be the cause for disease resistance. Cell death may result in production of toxic substances by activating enzymes which may release aglycones (phenolics) from glycosides (non-toxic phenol + sugar complex)

toxic phenol + sugar complex) which will be highly toxic to the pathogen. The phytoalexins may be produced by healthy cells adjacent to the dying cells.

In some cases, cell death may occur after the fungal death and in some other cases hypersensitive cell death may not be related to disease resistance at all. Thus hypersensitivity is not a rule but at least in few cases, hypersensitivity leads to disease resistance.

REFERENCES

1. Arrigoni, O.R., R. Arrigoni-Lisa and G. Calabrese, 1976. Ascorbic acid as a factor controlling the development of cyanide - insensitive respiration. Science, **194** : 332-333
2. Ayers, A.R., B. Valent, J. Ebel and P. Albersheim, 1976. Host-pathogen interactions. XI. Composition and structure of wall-released elicotor fractions. Plant Physiol, **57** : 766-774
3. Bailey, J.A., P.M. Rowel and G.M. Arnold, 1980. The temporal relationship between host cell death, phytoalexin accumulation and fungal inhibition during hypersensitive reactions of *Phaseolus vulgaris* to *Colletotrichum lindemuthianum*. Physiol. Plant Path, **17**: 329-339
4. Bird, P.M. and J.P. Ride, 1981. The resistance of wheat to *Septoria nodorum* : Fungal development in relation to host lignification. Physiol. Plant Path, **19**: 289-299
5. Chakravorty, A.K. and M. Shaw, 1977. The role of RNA in host-parasite specificity. Annu. Rev. Phytopathol, **15**: 136-151
6. Daly, J.M. and I. Uritani, 1979. Recognition and specificity in plant host-parasite interactions. Japan Scientific Societies Press, Tokyo and University Park Press, Baltimore
7. Hadwiger, L. A. and J.M. Beckman, 1980. Chitosan as a component of pea -*Fusarium solani* interactions. Plant Physiol, **66**: 205-211
8. Harborne, J.B., 1964. Biochemistry of phenolic compounds. Academic Press, London, p. 618
9. Health, M.C. 1976. Hypersensitivity, the cause or the consequence of rust resistance. Phytopathology, **66**: 935-936
10. Ingham, J.L., 1972. Phytoalexins and other natural products as factors in plant disease resistance. Bot. Rev., 38: 343-424
11. Ingram, D.S., 1978. Cell death and resistance to biotrophs. Ann. Appl. Biol., **89** : 291-295
12. Kassanis, B., 1980. Therapy of virus-infected plants and the active defense mechanism. Outlook Agriculture, 10: 288-292
13. Keen, N.T. and M. Legrand, 1980. Surface glycoproteins: evidence that they may function as the race specific phytoalexin elicitors of *Phytophthora megasperma* f. sp. *glycinea.* Physiol. Plant Pathol, **17**: 175-192
14. Kiraly, Z., B. Barna and T. Ersek, 1972. Hypersensitivity as a consequence, not the cause of plant resistance to infection. Nature, **239**: 456-458
15. Kosuge, T. 1969. The role of phenolics in host response to infection. Annu. Rev. Phytopathol, **7**: 195-222
16. Loebenstein, G. 1972. Localization and induced resistance in virus-infected plants. Annu. Rev. Phytopathol, 10: 177-206
17. Mapson, L. W. 1958. Ascorbic acid metabolism in plants. Annu. Rev. Plant Physiol, **9**: 118-145
18. Paxton, J.D., 1981. Phytoalexins - A working redefinition. Phytopath Z., **101**: 106-109
19. Sarkanen, K.V. and C.H. Ludwin, 1971. Lignins: Occurrence, formation, structure and reactions. Wiley-Interscience, New York, p. 916
20. Sequeria, L., 1978. Lectins and their role in host-pathogen specificity. Annu. Rev. Phytopathol, 16: 453-481
21. Vance, C.P., T.K. Kirk and R.T. Sherwood, 1980. Lignification as a mechanism of disease resistance. Annu. Rev. Phytopathol, **18**: 259-288

22. Vidhyasekaran, P., 1972. Nitrogen metabolism of ragi plants in relation to helminthosporiose disease resistance. Phytopath. Z. **74**: 223-235

23. Vidhyasekaran, P., 1974. Possible role of sugars in restriction of lesion development in finger millet leaves infected with *Helminthosporium tetramera.* Physiol. Plant Pathol, **4**: 457-467

24. Vidhyasekaran, P., 1974. Carbohydrate metabolism of ragi plants infected by *Helminthosporium nodulosum.* Phytopath. Z., **79**: 130-141

25. Vidhyasekaran, P., 1974. Finger millet helminthosporiose, a low sugar disease. Z. Pflkrankh. Pflanzen, **81**: 28-38

26. Vidhyasekaran, P., 1975. Role of the auxin-phenol complex in finger millet (*Eleusine coracana)* resistance to helminthosporiose. Phytopath. Z., 82: 89-99

27. Vidhysekaran, P., 1983. Host-pathogen specificity. In: A. Hussain, Kishan Singh, B.P. Singh and V.P. Agnihotri (eds.) Recent advances in Plant Pathology. Print House (India), pp. 287-296

28. Vidhyasekaran, P., 1988a. Physiology of disease resistance in plants. Vol. I, CRC Press, Florida, U.S.A.

29. Vidhyasekaran, P. 1988b. Physiology of disease resistance in plants. Vol. II. CRC Press, Florida, U.S.A.

30. Walker, J.C. and M.A. Stahmann, 1955. Chemical nature of disease resistance. Annu. Rev. Plant Physiol, **6**: 361-366

31. Wood, R. K. S., 1981. Active defense mechanisms in plants. Plenum Press, New York.

13

DISEASE DEVELOPMENT — ENVIRONMENT

Disease development depends upon three factors viz., host, pathogen and environment.

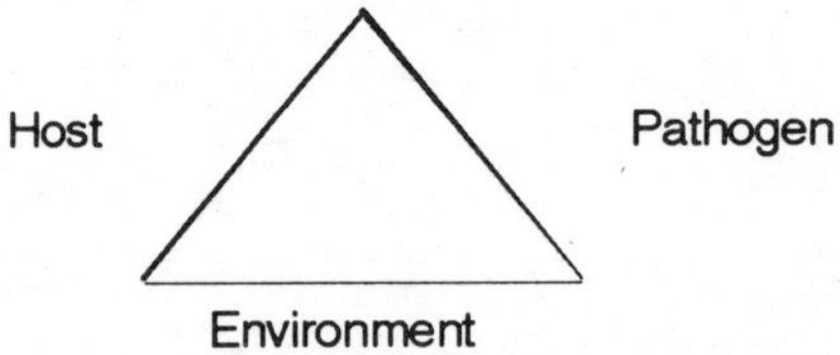

The host should be susceptible and the pathogen should be virulent to cause disease. When both the factors are present, environment plays an important role in disease development. Several environmental factors contribute to the disease development by altering host metabolism and/or virulence of the pathogen. The various environmental factors favourable for disease development are discussed in this chapter.

TEMPERATURE

Powdery mildew diseases are favoured by high temperature. Bean, greengram and blackgram powdery mildews *(Erysiphe cichoracearum)* are severe in summer. Root rots caused by *Macrophomina phaseolina* are also severe at high temperatures (30-35°C). Tobacco mosaic virus becomes severe on bean at high temperature.

Rice blast is a low-temperature disease. When the night temperature falls below 20°C, rice blast *(Pyricularia oryzae)* becomes severe. *Verticillium* wilt of cotton is severe at low temperatures. Loose smut of wheat is severe when the temperature is around 19-20°C. If the temperature is high (29-30°C), no disease develops.

Effect of Temperature on Pathogens

Temperature may favour the growth of pathogen which results in severe disease incidence. The optimum temperature for growth of rice bacterial blight pathogen, *Xanthomonas campestris* pv. *oryzae* is 25-30°C. The disease incidence is also favoured by temperatures above 25°C. Symptoms do not appear at temperatures below 20°C.

Puccinia striiformis, the wheat stripe rust pathogen does not tolerate high temperature and both spores and mycelium grow best around 11°C. The disease is practically absent in the seasons when the temperature is above 25°C.

Predisposing Effect of Temperature on the Host

Rice blast is severe under low night temperature conditions. At 20°C L-glutamine synthesis is high in rice plants and glutamine stimulates the spore germination of *Pyricularia oryzae,* the blast pathogen. At low temperature nitrogen content increases resulting in reduction in phenolics synthesis. Phenolics are fungitoxic and less phenolics content favours the disease incidence.

The optimum temperature for growth of the tobacco root rot pathogen, *Thielaviopsis basicola* in culture is 28°C. But it causes the greatest damage to tobacco at a temperature of 17° to 23°C at which temperatures the fungus does not grow well in culture. The increased disease incidence is due to poor growth of the host at low temperatures. The weakened host is affected severely by the pathogen.

HUMIDITY

Humidity is very much important in the development of diseases. Humidity induces sporulation in fungi. The sugarcane downy mildew fungus, *Peronosclerospora sacchari* produces more conidia when the relative humidity increases from 86 to 95% (4). Maximum sporulation of sorghum downy mildew fungus *(Peronosclerospora sorghi)* takes place at 100% relative humidity. At 80 per cent relative humidity and below, no conidia are produced.

Humidity alters the host physiology also. Ragi plants under high humid conditions contain less sugars and low sugar content favours *Helminthosporium nodulosum* leaf blight incidence (7).

DEW

Dew favours sporulation of many fungal pathogens. *Peronospora philippiensis,* the sorghum downy mildew fungus produces conidia when the leaves are wet for at least 4-5 hours. Dew has to be present for sporulation to occur (1). Length of dew period is very much important in rice blast *(Pyricularia oryzae)* disease development. Dew period of 6 to 8 hrs. highly favours the disease. Dew favours conidial germination of *P. oryzae.*

RAIN

Dispersal of spores occurs due to rain. In case of bajra downy mildew *(Sclerospora graminicola),* sporangia are dispersed up to 340 metres during the rainy season. But during dry season the dispersal is only up to 80 meters. Frequent drizzling rains favour bajra ergot *(Claviceps fusiformis)* development (6). Rice bacterial blight *(Xanthomonas campestris* pv. *oryzae)* becomes severe after the occurrence of heavy rain. Rain disperses the bacteria and causes epidemic.

WIND

Heavy wind causes severe injury in rice leaves. The wound injury favours entry of the bacterium *(X. campestris* pv. *oryzae)* into the host leaves. After heavy wind the rice bacterial blight becomes severe.

LIGHT

Light is known to influence sporulation of fungal pathogens. Increasing light duration increases sporangial production in case of bajra downy mildew fungus *(Sclerospora graminicola).* But sporangial liberation occurs under darkness. Tobacco necrosis virus infection in tobacco is severe under low light intensity. Under high light intensity the disease is negligible and only a few lesions appear. Shading increases brown spot *(Helminthosporium oryzae)* incidence. Shade will decrease sugar content resulting in less phenolics. Less phenolics lead to more disease.

COMPLEXITY OF WEATHER FACTORS

We have seen that many weather factors contribute for the disease incidence. Single weather factor alone may not contribute for the disease incidence in many cases. For example, a combination of dew period and low night temperature favour the rice blast incidence while minimum temperature and relative humidity favour tikka (late) leaf spot incidence in groundunt. Both rainfall and temperature are considered to be key factors in the outbreak of late blight of potato. A combination of several favourable weather factors will cause epidemic.

SOIL ENVIRONMENT

Moisture

Soil moisture determines the outbreak of soil-borne diseases. Soil moisture depends upon the amount of water present in the soil. Water is soil in described in terms of its protential energy. It is a called water potential which is defined as the capacity to do work relative to the work capacity of pure, free water. Water with a work capacity greater than that of pure free water (e.g., water under pressure or elevated above the reference point) is at a positive water potential (+ bars; 1 bar = 100 joules/kg). Water with work capacity less than pure, free water (eg., with salts or sugar dissolved, or absorbed to a matrix such as soil) is at a negative water potential (-bars). The lower the water potential, the greater the water stress. There are two types of water potential in soil. Matric water protential depends upon soil water content while osmotic water potential depends upon salts in soil water. Soil water potential is the sum of matric potential and osmotic potential.

Phytophthora cinnamomi grows optimally at -5 bars or wetter and does not grow at all at water potentials below -50 bars. The damping off diseases caused by the pathogen are always severe in wet soils and absent in dry soils. Contrastingly *Fusarium solani* grows well at -10 to -30 bars and is prevented only if the water potential approaches or drops below-100 bars. The dry root rot diseases caused by *F. solani* are severe in dry soils and less or absent in wet soil (2). Potato scab caused by *Streptomyces scabies* is severe during dry period. In moist soil the disease is negligible.

Soil Temperature

Cotton crop is affected by two wilt pathogens. *Verticillium albo-atrum* causes severe wilt incidence when the soil temperature is low (about 20°C). In contrast, *Fusarium oxysporum* f. sp. *vasinfectum* causes severe wilt incidence when soil temperature increases to 32°C. *V. albo-atrum* grows well at low temperatures while *F. oxysporum* f. sp. *vasinfectum* grows well at high temperatures (about 30°C).

Soil pH

Cotton wilt caused by *F. oxysporum* f. sp. *vasinfectum* is severe at acid pH while it is almost absent in alkaline soil. Tomato wilt caused by *F. oxysporum* f. sp. *lycopersici* is also severe at low pH. In alkaline soil micronutrients like zinc and iron required for growth of the pathogen are not available. Hence, the pathogen does not grow well and cause disease.

REFERENCES

1. Dalmacio, G. and A.D. Reymundo, 1972. Spore density of *Sclerospora philippiensis* in relation to field temperature, relative humidity and downy mildew incidence. Philippine Phytopathology, **8**: 72-77
2. Cook, R.J. 1973. Influence of low plant and soil water potentials on diseases caused by soil-borne fungi. Phytopathology, **63**: 451-458
3. Prasada, R. and U.N. Verma. 1948. Studies on lentil rust, *Uromyces fabae*. Indian Phytopath, **1** : 142-146
4. Shekara Shetty, H. 1990. Basic Research on downy mildews of cereals and disease management. In: P. Vidhyasekaran (ed.) Basic Research for crop disease management, Daya Publishing House, Delhi, pp. 316-342
5. Subba Rao, K.V. and R.P. Thakur, 1983. *Tolyposporium penicillariae,* the causal agent of pearl millet smut. Trans. Br. Mycol. Soc., **81**: 597-603
6. Thakur, R.P. 1990. Basic Research on management of pearl millet diseases. In: P. Vidhyasekaran (ed.) Basic Research for Crop Disease Management. Daya Publishing House, Delhi, pp. 343-358
7. Vidhyasekaran, P. 1974. Role of humidity on the incidence of ragi helminthosporiose. Indian Phytopath, **27**: 242
8. Vidhyasekaran, P. 1990. Basic Research for Crop Disease Management. Daya Publishing House, Delhi, p. 410
9. Weston, W.H. Jr. 1923. Production and dispersal of conidia in the Philippines. Indian J. Agric. Sci., **48**: 537-539.

14

FORECASTING EPIDEMICS

The ultimate goal of the science of 'Plant Pathology' is a complete understanding of plant disease - host, pathogen and environment interactions - so that disease control can be obtained when necessary and economically warranted. Disease monitoring and forecasting systems are being developed for effective management of various diseases.

To develop a forecasting system, a complete knowledge on the factors affecting disease development should be available. The development of the disease depends upon:

(i) the amount of inoculum of the pathogen present at a particular time,
(ii) susceptible stage of the crop, and
(iii) weather factors favouring infection. Weather factors include the magnitude, duration, and relationships among humidity, temperature, rain fall, dew periods, radiation and air movement.

Forecasting systems can be classified into two systems:

(i) empirical system, and
(ii) fundamental system

Empirical Forecasting System

The system has been developed by studying and comparing historical records of disease occurrence and concurrent weather conditions in the same or approximately same locality. Such systems usually result in the formulation of 'rules'. The rules enumerate the specific meteorological conditions that must be fulfilled before disease development can take place.

Forecasting systems based on synoptic charts (the charts which display weather conditions as they exist simultaneously over a broad area) have been developed. For example, late blight of potato is forecasted in Europe if:

(i) open waves of maritime tropical air, and
(ii) stagnant or slow moving depressions giving lengthy period of wet, overcast weather, occur.

Fundamental Forecasting System

Predictive systems developed from data obtained from experiments in the laboratory or fielding regarding the relationships of biological and environmental conditions governing host-pathogen interactions are called fundamental or 'derived' or 'inductive' or logical systems.

Normally after developing empirical forecast system, the system is tested and modified to fit various

geographical and environmental conditions. Some questions arise during this type of analysis and these questions are answered under controlled experimental conditions. Then, the relationships are defined mathematically. Now the system becomes fundamental system.

Fundamental forecasting system is developed based on multiple regression analysis involving several independent biological and climatological variables to describe the progress of an epidemic.

In Germany, late blight of potato is forecasted based on stepwise multiple regression using 5 biometeorological parameters. The parameters used are as follows:

(i) Meteorological influence on sporangium production.
(ii) Meteorological influence on sporangial germination.
(iii) Meteorological influence on colonization.
(iv) Magnitude and duration of high humidity periods.
(v) Effect of dry period on disease development.

All the data are collected daily and analysed by a computer programme which produces output in the form of ratings accumulated from the average time of plant emergence. No disease forecast is made until 150 ratings are accumulated. When 150 ratings are observed, an alert is given that blight could occur within 15-40 days. When 270 ratings are accumulated blight is forecasted to occur within 15 days after this date. At this stage, a call for control measures is made.

Conditions for Successful Forecasting

1. When the number of parameters necessary to accurately predict infection increases the probability of making accurate forecast decreases. Less number of parameters should be used.
2. The infection rate of the pathogen should not exhibit variability each year. Then only the rate can be used to predict future epidemic development.
3. Forecasting system should be developed only for the disease which is not controlled by genetic resistance and which can completely destroy a major food crop. It should also be amenable for chemical control.

EXAMPLES OF WELL DEVELOPED FORECASTING SYSTEM

Tikka Leaf Spot of Groundnut

This disease causes heavy losses and can be effectively controlled by spraying fungicides. A technique for forecasting the disease has been developed in U.S.A.

When groundnut foliage remains wet for a period greater than or equal to 10 hrs. and the minimum temperature is 21°C or higher for two consecutive days or nights, the disease development is forecasted (3).

A computer programme has been developed and widely used in U.S.A. The data on hours per day with relative humidity (RH) of 95% and above and the minimum temperature (T) during the RH observation period for the previous 5 days are fed to the computer. Calculations are rounded to whole numbers. The T/RH index for each of the five days is calculated. For example, when hours of the RH 95% equal 10 and the minimum temperature during the period equals 21.1°C, the T/RH index is 2.0. The T/RH indices for days 4 and 5 are summed. If the total index exceeds, 4, disease is forecasted. If the index is 3 or less no disease is forecasted. This computer forecast is accurate to a great extent in forecasting the disease (6).

Late Blight of Potato

Late blight also causes heavy losses in potato crop and can be effectively controlled by spraying fungicides. In U.S.A. a good forecasting programme has been developed.

The initial appearance of late blight is forecasted 7-14 days after the occurrence of 10 consecutive blight-favourable days. A day is considered to be blight - favourable when the 5 days average temperature is 25.5°C and the total rainfall for the last 10-days period is more than 3.0 cm. This system is known as the "moving graph" system because it is always based on the average temperature and the total rainfall for the last 5 and 10 days, respectively.

A computerized version (blitecast) has also been developed in U.S.A. (5).

REFERENCES

1. Berger, R.D. 1977. Application of epidemiological principles to achieve plant disease control. Annu. Rev. Phytopathol, **15**: 165-183
2. Hyre, R.A. 1954. Progress in forecasting late blight of potato and tomato. Plant Dis. Reptr., **38**: 245-253
3. Jensen, R.E. and L.W. Boyle., 1966. A technique for forecasting leafspot on plants. Plant Dis. Rept.r, **50**: 810-814
4. Krause, R.A. and L.B. Massie. 1975. Predictive systems: Modern approaches to disease control. Annu. Rev. Phytopathol, **15**: 165-183
5. Krause, R.A., L.B. Massie and R.A. Hyre. 1975. Blitecast: A computerized forecast of potato late blight. Plant Dis. Reptr., **59**: 95-98
6. Parvin, D.W., D.H. Smith and F.L. Crosby. 1976. Development and evaluation of a computerised forecasting method for *Cercospora* leaf spot of plants. Phytopathology, **64**: 385-388
7. Tarr, S.A.J., 1972. The principles of Plant Pathology, Winchester Press, p. 625
8. Wallin, J.R. 1951. Forecasting tomato and potato late blight in the north central region. Phytopathology, **41**: 37 (Abstr.).

15

DISEASE MANAGEMENT — QUARANTINE

The best way to control diseases is to exclude them from the area of sphere in which the host plants are being grown. This method of exclusion enforced through certain legal measures is commonly known as quarantine. Quarantine prevents the entry of new diseases into a new area. In the last century grapevine downy mildew *(Plasmopara viticola)* was introduced into France from U.S.A. The most devastating disease in coffee, coffee rust *(Hemileia vastatrix)* was introduced to India from Srilanka in 1876. Fire blight of apple *(Erwinia amylovora)* was introduced to India from England in 1940. Bunchy top of banana was introduced from Sri Lanka in 1940 to South India. The golden nematode of potato was introduced from Europe to India in 1960s. Groundnut rust has been introduced from U.S.A. to India in 1970s.

All those introduced diseases have shattered the economy of the countries into which they were introduced. Hence an International Plant Protection Convention was held in 1951 under the guidance of FAO (Food and Agricultural Organization of United Nations). The resolution said that member countries should cooperate with one another in the prevention of introduction and spread of plant and plant products carrying destructive diseases across international borders. Under this convention regional bodies have been set up. One such body is for South East Asia and the Pacific region. For Europe, European and Mediterranean Plant Protection Organization (EPPO) has been formed.

Quarantine in India

In India, Destructive Insects and Pests Act, was enacted in 1914. Directorate of Plant Protection and Quarantine was established in Delhi. It is under the Ministry of Agriculture, Government of India under the overall supervision of Plant Protection Adviser. Quarantine stations have been established in airports (Bombay, New Delhi, Madras, Calcutta, Amritsar, Tiruchi and Trivandrum), seaports (Bombay, Madras, Calcutta, Cochin, Tuticorin, Visakhapatnam, Rameshwaram and Bhavnagar), land frontiers of Attari-wagha border (Amritsar, Kallimpong, etc.) and in Railway stations at Attari and Amritsar.

Plants and plant products to pass through these quarantines should be accompanied by a phytosanitary certificate, as per international convention, issued by the competent authority of the exporting countries. All these products will be inspected and fumigated at these quarantine stations if necessary. To export the plants and plant products from India, State Plant Pathologists have been empowered to issue phytosanitary certificates.

Embargo

Some of the plant materials have been totally banned for import into India irrespective of any certificate.

Potatoes cannot be imported to India from any wart disease *(Synchytrium endobioticum)* and/or golden nematode-infested areas. Unginned cotton cannot be imported. Sugarcane cuttings from Australia, Fiji and Philippines cannot be imported. Import of cocoa from Africa and Sri Lanka is totally banned. Rubber plant and seeds from South America and West Indies and sunflower seeds from Argentina and Peru are totally banned.

Domestic Quarantine

Within India, domestic quarantines have been established. To prevent the spread of bunchy top of banana, an act has been promulgated which prevents the movement of suckers, stems and materials of banana plant used for packing and wrapping excepting fruits from the states of Assam, Kerala, West Bengal and Orissa to other states. To prevent the spread of banana mosaic, movement of banana suckers from Gujarat and Maharastra to other parts of India is prohibited. Potato wart *(Synchytrium endobioticum)* is prevalent in Darjeeling district of West Bengal and hence movement of potatoes from West Bengal to other states is prohibited. Similarly golden nematode is prevalent in the Nilgiris district of Tamil Nadu and movement of potatoes from this district to other parts of India is prohibited.

REFERENCE

1. Chattopadhyay, S.B. 1985. Principles and Procedure of Plant Protection. Oxford & IBH Publishing Co., New Delhi, p. 547.

16

DISEASE MANAGEMENT - CULTURAL PRACTICES

Every operation of cultural practices may affect the disease incidence. Hence by proper cultural practices many diseases can be controlled. Some of the examples are given in this chapter.

1. Seeds

Seeds carry many fungi, bacteria and viruses and may introduce these pathogens into the field. Loose smut of wheat *(Ustilago tritici),* black arm of cotton *(Xanthomonas campestris* pv. *malvacearum)* and leaf crinkle of blackgram (virus) are introduced into the field through seed. Hence, the infected seeds should not be used for sowing.

2. Vegetative Propagating Material

Banana suckers may carry the bunchy top virus. Hence suckers from diseased areas should not be used. Sugarcane setts may carry red rot pathogen (*Colletotrichum falcatum).* Hence sugarcane setts from diseased areas should not be used for planting.

3. Irrigation and Drainage Water

Soil-borne pathogens may be carried from one field to another field through irrigation or drainage water. Hence care should be taken not to irrigate a healthy field using drainage water from an infected field. Bacterial blight of rice *(Xanthomonas campestris* pv. *oryzae)* spreads mostly through irrigation and drainage water (16).

4. Plant Debris

Some of the pathogens survive in plant debris; eg. stem rot of rice *(Sclerotium oryzae)* and black arm of cotton *(X campestris* pv. *malvacearum).* Hence plant debris should be burnt or removed.

5. Equipment and Men

Pruning implements, intercultivators, etc., may carry some pathogens; eg. tobacco mosaic virus in tobacco field. Workers also may carry the virus from infected field to the healthy field. Hence proper sanitation is needed.

6. Weeds

Weeds serve as alternate hosts for many pathogens; eg. rice tungro virus, rice blast, etc. Hence clean cultivation is essential for control of diseases.

7. Volunteer Plants

Self-sown volunteer plants may carry infection which may serve as source of infection for the succeeding crop; eg. cotton black arm. Volunteer plants should be destroyed.

8. Ratoons

Ratoon crops normally carry many pathogens; eg. sugarcane grassy stunt. Ratoon crops should be avoided as far as possible.

9. Roguing Infected Plants

The infected plants should be rogued out to reduce the inoculum in the field. The smutted whip should be removed to reduce sugarcane smut incidence. The virus-infected plants should be removed to reduce the virus inoculum. Yellow mosaic of greengram can be reduced by roguing infected plants.

10. Pruning of Infected Twigs and Branches

The removal of infected twigs may reduce disease incidence. Pruning of citrus canker affected branches reduces the disease incidence.

11. Removal of Affected Portion

Surgery is carried out to remove affected woody portion to control citrus gummosis caused by *Phytophthora citrophthora*.

12. Burning Crop Residue

The crop residue is burnt to avoid carry-over of inoculum in case of rice sheath blight *(Rhizoctonia solani)* and rice stem rot *(Sclerotium oryzae)*.

13. Solar Heating

When the soil is covered with polyethylene sheets during hot seasons, soil temperature increased and eliminated *Verticillium dahliae* and *Fusarium oxysporum* f. sp. *lycopersici* in tomato field. High temperature favours growth of antagonistic fungi also (10).

14. Flooding the Field

Fusarium oxysporum f. sp. *cubense* does not survive well in flooded field. Hence *Fusarium* wilt of banana can be controlled by flooding the field for long time (24). Tobacco black shank *(Phytophthora parasitica* var. *nicotianae)* and cotton wilt *(Verticillium dahliae)* can also be controlled by flooding the field.

15. Fallowing

By fallowing, *F. oxysporum specific* strains can be reduced as they do not survive in soil for long time. But fallowing will not reduce *Rhizoctonia solani* infection.

16. Date of Sowing

By adjusting the time of sowing some of the diseases can be controlled; eg. early sowing of pearl millet will reduce the ergot disease. Avoiding wet seasons for planting of rice will eliminate rice blast.

17. Avoidance of Monoculture

Continuous cultivation of rice crop may result in the appearance of new pathotypes of *Pyricularia oryzae* the rice blast pathogen.

18. Multiple Cropping

Taller crops can be sown to protect a crop of lesser height from virus vectors. The insects may land at the taller crops and the dwarf crop may escape from virus diseases transmitted by those insects.

19. Intercropping

Intercrops should be properly chosen so that all the crops should not have any common pathogen. For example, *Rhizoctonia solani* has wide host range and hence common hosts of the pathogen should not be grown as intercrops.

20. Trap Crop

Tagetes is a trap crop for many nematodes and can be grown to reduce nematode infection. Fodder sorghum can be raised to reduce downy mildew of sorghum.

21. Tillage

Ploughing of rice residue to a depth of 25-30 cm renders a high percentage of inoculum of *Sclerotium oryzae* unavailable for infection of rice subsequently sown with flooding.

22. Soil Amendment

Potato black scurf *(Rhizoctonia solani)* is less in soil amended with wheat straw (5). Pea root not *(Aphanomyces euteiches)* is reduced when cruciferous plant residue is incorporated in soil. Lucerne meal and barley straw reduce root rot of cotton caused by *Macrophomina phaseolina.*

23. Seed Rate and Spacing

Closer spacing favours many air-borne diseases because of high humidity in the crop canopy. Tikka leaf spot of groundnut *(Cercospora arachidis)* and blister blight of tea *(Exobasidium vexans)* are more in dense canopy. Damping off caused by *Pythium and Phytophthora* increases in crops when higher seed rate was used.

Crowded stands may reduce some systemic diseases. Cotton wilt caused by *Verticillium albo-atrum* will be less in closely planted crop if the fungal inoculum is less in the soil. Similarly closer spacing of rice reduces rice tungro infection particularly when vector population is less.

24. Irrigation

Potato scab *(Streptomyces scabies)* is prevented by maintaining soil water near field capacity during the tuber formation. High soil moisture content encourages antagonistic bacterial population. Irrigation controls *Macrophomina phaseolina* infection in potato and cotton (3).

25. Nitrogenous Fertilizers

(a) *Dosage:* Increased application of nitrogen increases many diseases. Rice blast becomes severe in nitrogen applied fields (18).

(b) *Method of application:* Split application reduces blast and bacterial blight of rice. Contrarily late application of nitrogen increases wheat leaf blotch *(Septoria nodorum)* and powdery mildew *(Erysiphe graminis)* (13).

(c) *Type of N fertilizer:* Some diseases are favoured by ammoniacal form of nitrogen while others are favoured by nitrate form of nitrogen. Wilts *(Fusarium* spp.) and roots *(Rhizoctonia* spp.) are favoured by ammoniacal nitrogen while *Verticillium* wilts and root rots due to *Pythium* spp. are favoured by nitrate nitrogen (23).

In rice, blast disease is favoured by ammoniacal nitrogen while brown spot (*Helminthosporium orzae)* is favoured by nitrate nitrogen. In maize leaf spot caused by *Helminthosporium turcicum* is favoured by ammonical nitrogen while stalk rot *(Diplodia maydis)* is favoured by nitrate nitrogen. In wheat, sharp eyespot *(Rhizoctonia solani)* is favoured by ammonical nitrogen while stem rust *(Puccinia graminis tritici)* is favoured by nitrate nitrogen. In potato, wilt caused by *(Verticillium albo-atrum* and scab caused by *Streptomyces scabies* are favoured by nitrate nitrogen while ammoniacal nitrogen increases black scurf *(R. solani)* (8).

26. Phosphorous

Repeated application of phosphatic fertilizers delays the onset and lessens the severity of take-all disease of barley *(Gaeumannomyces graminis)* (14).

27. Potassium

Potassium application reduces the disease incidence in many cases probably by increasing phenolics synthesis in plants (20).

28. Calcium

Calcium application reduces bean root rot caused by *Rhizoctonia solani* probably by altering pectin metabolism of the host (1).

29. Molybdenum

Application of molybdenum reduces infection of potato tubers by *Phytophthora infestans* (17) and also diminishes incidence of *Ascochyta* blight on beans and peas.

30. Manganese

Manganese reduces late blight of potato (19).

31. Iron

Ferric chloride controls rice brown spot (18).

32. Silicon

Silicon application reduces rice blast *(Pyricularia oryzae)* (18).

REFERENCES

1. Baker, K.F. and R.J. Cook. 1974. Biological control of plant pathogens. W.H. Freeman and Co., San Francisco, p. 433
2. Butterfield, E.J., J.E. De Van and R.H. Garber. 1978. The influence of several crop sequences on the incidence of *Verticillium* wilt of cotton and on the population of *Verticillium* dahliae in field soil. Phytopathology, **68**: 1217-1220

3. Cook, R.J. and R.J. Papendick, 1972. Influence of water potential of soils and plants on root disease. Annu. Rev. Phytopathol, **10**: 349-374
4. Dasgupta, M.K. and S.B. Chattopadhyay, 1977. Effect of different doses of N and P on the susceptibility of rice to brown spot caused by *Helminthosporium oryzae*. Z. Pflanzenkrankh. Pflanzenschutz, **84**: 276-285
5. Gudmestad, N.C., J.E. Huguelet and R.T. Zink, 1978. The effect of cultural practices and straw incorporation into the soil on *Rhizoctonia* disease of potato. Plant Dis. Reptr., **62**: 985-989
6. Henis, Y. and I. Chet. 1967. Mode of action of ammonia on *Sclerotium rolfsii*. Phytopathology, **57**: 425-427
7. Hough, A., N.J. Mulder and J.M. La Granage, 1979. Heat treatment for the control of *Phytophthora* gummosis in citrus. Plant Dis. Rep., 63: 40-43
8. Huber, D.M. and R.D. Watson, 1974. Nitrogen form and plant disease. Annu. Rev. Phytopathol, **12**: 139-165
9. Jones, J.P. and S.S. Woltz, 1972. Effect of soil pH and micronutrient amendments on *Verticillium* and *Fusarium* wilt of tomato. Plant Dis. Reptr., **56**: 151-153
10. Katan, J. 1980. Solar pasteurization of soils for disease control: Status and Prospects. Plant Dis., **64**: 445-454
11. King, S.B. and O.J. Webster, 1970. Downy mildew of sorghum in Nigeria. Indian Phytopathol, **23**: 342-349
12. Kraft, J.M. and R.A. Giles. 1979. Increasing green pea yields with root rot resistance and subsoiling. In: Schippers, B. and W. Gams (eds.) Soil-borne Plant Pathogens. Academic Press, New York, pp. 407-413
13. Last, F.T. 1954. The effect of time of application of nitrogenous fertilizers on powdery mildew of wheat. Ann. Appl. Biol., **41**: 381-392
14. Mattingly, G.E.G. and D.B. 1974. Slope, Phosphate fertilizer and 'take all' disease of wheat and barley. J. Sci. Food Agric. **28**: 658-659
15. Mayee, C.D., B.K. Rattan and H.N. Shahi, 1976. Agro-practices in the incidence of stem rot disease of rice in the Punjab. Indian J. Mycol. and Plant Pathol, **6**: 166-170
16. Mizukami, T. and S. Wakimoto. 1969. Epidemiology and control of bacterial leaf blight of rice. Annu. Rev. Phytopathol, **7**: 51-52
17. Mudich, A. 1967. Effect of trace elements bound to superphosphate on the resistance of potato tubers to *Phytophthora infestans* (Mont.) de Bary. Acta Phytopath. Hungaricae, **2**: 295-302
18. Ou, S.H. 1985. Rice Disease. CMI, Kew, Surrey, U.K., p. 380
19. Palti, J. 1981. Cultural practices and infectious crop diseases. Springer-Verlag, Berlin, p. 243
20. Perrenoud, S. 1977. Potassium and plant health. IPI Research Topics. In Potash Inst., Bern, p. 218
21. Putter, C.A.J. 1980. The management of epidemic level of endemic diseases under tropical subsistence farming conditions. In: Palti J. and J. Kranz (ed.) Comparative epidemiology. PUDOC, Wageningen, p. 93-103
22. Rotem, J. and J. Palti. 1969. Irrigation and plant diseases. Annu. Rev. Phytopathol, **7**: 267-288
23. Smiley, R.W. 1975. Forms of nitrogen and the pH in the root zone and their importance to root infections. In: Bruchl, G.W. (ed.) Biology and Control of soil-borne pathogens. Amer. Phytopathol. Soc. St. Paul, Minnesota, pp. 52-62.
24. Stover, R.H. 1955. Flood fallowing for eradication of *Fusarium oxysporum* f. *cubense*. III Effect of oxygen in fungus survival. Soil Sci., **80**: 397-412
25. Stover, R.H. 1956. Studies on *Fusarium* wilt of bananas. I. The behaviour of *F. oxysporum* f. *cubense* in different soils. Can. J. Bot., **34**: 927-942

17

DISEASE MANAGEMENT - USE OF RESISTANT CULTIVARS

Host plants are known to have genes for resistance to diseases. Several types of resistance have been reported in plants.

Vertical and Horizontal Resistance

Vertical resistance is otherwise called as specific resistance, race-specific resistance, qualitative resistance, true resistance, monogenic resistance, or major gene resistance. Horizontal resistance is otherwise called as non-specific resistance, race-non-specific resistance, quantitative resistance, general resistance, polygenic resistance, or minor gene resistance.

When a variety is more resistant to some race of a pathogen than to others, the resistance is called 'vertical'. Vertical resistance reduces the effective amount of initial inoculum from which the epidemic starts. A feature of vertical resistance is that the infection rate is as fast in the vertically resistant as the completely susceptible variety after the initial infection has occurred. In other words, the pathogen may mutate and cause severe disease in the vertically resistant variety. The vertical resistance may break quickly whenever new races are formed. Most vertical resistance apparently is controlled by single genes.

When host resistance is equally effective against all races of a pathogen - it is termed 'horizontal;'. The action of horizontal resistance is to slow down the epidemic after it has started. Horizontal resistance can vary from zero (no resistance at all) to 'practically absolute immunity'. Horizontal resistance is called field resistance and is controlled by several genes.

Vertical resistance is recommended for intensive agriculture while horizontal resistance is for subsistence agriculture.

Complete, Incomplete, Partial and Rate-reducing Resistance

When multiplication of the pathogen is totally prevented i.e. when the spore production is zero, the resistance is called complete resistance. In incomplete resistance some spore production is seen. Partial resistance is a form of incomplete resistance in which spore production is reduced even though the host plants are susceptible to infection. Partial resistance can also be called as rate-reducing resistance. The resistance that affects infection frequency (lesion number), latent period and spore production is called rate-reducing resistance. In the rate-reducing resistance, latent period may be more or colony and lesion size may be small or spore production may be less. Latent period is the crucial component determining the apparent infection rate when a large number of reproductive cycles of the pathogen are required to complete the epidemic. Many rate reducing resistance are of polygenic in nature. But single gene resistance also can act as rate-reducing resistance. In tomato single resistance gene confers rate-reducing resistance. The genes C_f 1 and C_f 3 allow flecks to develop with some sporulation of *Cladosporium fulvum.* In wheat Sr genes Sr 7b, Sr 8, and Sr 14 have an infection type 2 with some sporulation in case of *Puccinia graminis* f. sp. *tritici* infection.

Slow-rusting (slow development of rust) in wheat and slow blasting in rice are common examples of rate-reducing resistance.

Durable Resistance

Durable resistance to a disease is resistance that remains effective during its prolonged and widespread use in an environment favourable to the disease. The test for durable resistance includes two elements viz. time (long) and area (large).

Durable resistance is found in both single gene resistance and polygenic resistance. Wheat cultivar 'Timgalen' which possesses Sr 5, Sr 6, Sr 8 and Sr 36 genes was released in 1967 and it is still resistant to stem rust in Australia. Races with pathogenicity for Sr 5, Sr 6 and Sr 8 are common in Australia and it appears that continuing resistance is mainly or solely due to Sr 36. This durable resistance is due to combination of genes each of which is known to be race-specific and none of which provided durable resistance when used separately or in other combinations.

Wheat cultivar 'Selkrik' is cultivated in Canada for many years showing resistance to stem rust. It contains Sr 2, Sr 6, Sr 7b, Sr 9d and Sr 23 genes. Sr 2 has incomplete resistance or slow rusting. Sr 2 is mainly responsible for durable resistance.

Tolerance

Tolerance is the ability of the host to endure the presence of the pathogen, and can be expressed by less severe (reduced) disease symptoms and/or less (reduced) damage. Tolerance may be due to different types of resistance genes.

Seedling Resistance and Adult Plant Resistance

In some cultivars only seedlings show resistance to a disease and in some others adult plants alone show resistance. Seedling resistance has been reported in some rice varieties against bacterial blight. Adult plant resistance has been reported in wheat varieties having Sr 2 gene for resistance to stem rust.

Resistance is Governed by Single Gene

Nelson (1978) reviewed the genetics of disease resistance. According to him there is only one type of resistance gene. Such genes are race-specific when alone or in small numbers. They provide race non-specific resistance when combined in large numbers. Wheat variety Frontana possesses as many as 5 stem rust resistant genes (Sr 3, Sr 6, Sr 9 and Sr 11). This type of 'pyramiding' genes (adding resistance genes) give an enduring control against wheat rust. It is also possible to combine various vertical resistance genes according resistance to diverse pests and diseases. It is called 'multiple disease resistance' or 'multiple resistance'.

Gene Blocks

Some of the resistant genes are closely linked. For example brown rust resistant gene (Lr 19) is closely linked with stem rust gene (Sr 25) in wheat varieties. This type of 'Gene blocks' can accord durable resistance.

Strong Gene and Weak Gene

Some genes accord resistance to almost all the races. The genes Lr 9 and Lr 19 derived from *Aegilops umbellulata* and *Agropyron* spp. are universally resistant to wheat brown rust. These genes are called strong genes. A strong gene accords resistance against a wide spectrum of races and lasts longer. In due course, the resistance breaks.

In contrast to the strong genes, Lr 9 and Lr 19, the genes Lr 14a, Lr 13, Lr 2a accord resistance to just a few races. They are called weak genes. Weak genes do not provide adequate resistance. The pathogen overcomes it quickly.

Aggressive Races and Virulent Races

Similar to strong gene and weak gene in host, in pathogens aggressive and virulent races may exist. Aggressive races differ in relative virulence. Some races may be highly virulent while others may be less virulent causing less damage to the crop. Virulent races are those which are pathogenic in some varieties while non-pathogenic in other varieties.

Multiline Cultivars

Several varieties are mixed and grown to control diseases. 'Clean crop' multilines are those in which each variety accords a broad spectrum of resistance. "Dirty crop" multilines are those in which a susceptible variety is purposefully introduced in the multiline mixture. The susceptible variety will reduce the selection pressure on the pathogen and retard the appearance of new races.

A mixture usually yields at least as well as the mean of their components, often more so, and sometimes exceed the highest component. It is uncommon for a mixture to yield less than the mean of its compounds, and rare for it to be as low or lower than the worst component.

Host mixtures may restrict the spread of a disease considerably relative to the mean of their components. Reduction upto 80% in powdery mildew infection compared with the mean disease level of the component grown as pure stands has been reported in barley. Reduction in disease incidence may be due to the decrease in the spatial density of susceptible plants. It limits the amount of susceptible tissue in a given area. It reduces the probability of survival of spores that leave each parent lesion. Spore density declines exponentially along a gradient from the source.

Varietal mixture may reduce the disease incidence due to the barrier effect provided by resistant plants that fill the space between the susceptible lines.

The third cause for reduced disease incidence may be due to the resistance induced by incompatible pathogenic spores. Induced resistance is always greater than induced susceptibility. Induced resistance in the varietal mixtures will be great.

In the varietal mixture slower adaptation of the pathogen is common compared to single variety cultivation.

Management of Diseases Using Resistant Cultivars

Several types of resistant cultivars are now available. There is no answer to the question, 'which type of resistance is preferred?' Vertical resistance can be preferred against pathogens which do not have many races or in the host which is not cultivated throughout the year. In the latter case there will not be continuous selection pressure on the pathogen to develop new races. Strong gene is preferred than weak gene for resistance. Durable resistance is preferred. Horizontal resistance will be useful in subsistence agriculture. Rate-reducing resistance will be useful against diseases which develop very fast such as rice blast and wheat rust. Varietal mixtures in wheat and rice will be of much use. Judicial management of diseases by employing different resistant varieties is the best and cheap method of disease control.

REFERENCES

1. Crill, P. 1977. An assessment of stabilizing selection in crop variety improvement. Annu. Rev. Phytopathol, 15: 185-202
2. Johnson, R. 1984. A critical analysis of durable resistance. Annu. Rev. Phytopathol, 22: 309-330
3. Nelson, R.R. 1978. Genetics of horizontal resistance to plant diseases. Annu. Rev. Pahytopathol, 16: 359-378
4. Wolfe, M.S. 1985. The current status and prospects of multiline cultivars and variety mixtures for disease resistance. Aninu. Rev. Phytopathol. 23 : 251-273.

18

DISEASE MANAGEMENT — BIOLOGICAL CONTROL

Biological control of diseases includes management of diseases by organisms including plants, but excluding man. Biological control can be achieved by:

(i) using microorganisms,
(ii) changing growth conditions of the plants (cultural practices); and
(iii) using resistant varieties.

Biological control using microorganisms has been developed both against soil-borne diseases and air-borne diseases.

I. Soil-borne Diseases

Rhizoctonia, Pythium, Sclerotium, and *Fusarium* are the important soil-borne pathogens which cause severe diseases in crops. They survive in soil for long time and in the same soil many antagonistic microorganisms are also present. The important antagonistic fungi are *Trichoderma* and *Gliocladium* species. They are widely distributed all over the world and occur in all soil. *Trichoderma* is found on well, decomposed organic matter, root surfaces of various plants, on decaying bark and on sclerotia of other fungi.

Some of the terms which are often used by the Plant Pathologists engaged in the biological control studies have to be defined for easy understanding of the subject.

Soil Fungistasis

When the fungal resting structures (sclerotia, chlamydospores, oospores, conidia, etc.) are added to the soil they do not germinate and grow in the soil. This condition is called soil fungistasis. Soil fungistasis is a widespread natural soil toxicity, which inhibits propagule germination. The cause for soil fungistasis is complex. Fungistasis is controlled by a complex balance of stimulators and inhibitors in soil micro-environments. The fungistasis in soil can be annulled (removed) by addition of nutrients to soil (crop residues, root-exudates and various amendments).

Suppressive Soils

In some soils, soil-borne pathogens do not survive and cause diseases. When the spores germinate after the annulment of fungistasis, the hypha gets lysed or again dormant spores are formed. This kind of soils are called suppressive soils. Suppressiveness has been attributed due to several factors such as:

(i) Antagonistic effects of associated microflora,
(ii) Faster multiplication of bacteria in suppressive soils,
(iii) Elevated fungistasis to fungal propagules,
(iv) Microbial competition for food,
(v) Ethylene production in soil, and
(vi) Unidentified biological factors that affect formation and persistence of resulting spores in soil.

Pseudomonas, Chromobacterium and *Arthrobacter* are the most important bacteria which are responsible for suppressive soil.

Hyperparasites

Hyperparasites are fungi which parasitize other fungi attacking both hyphae and sporulating structures. *Darluca, Tuberculina, Verticillium* and *Trichoderma* are the important hyperparasites which parasitize uredo and teleutosori of numerous rust species, conidia of powdery mildews and hyphae of *Rhizoctonia* species.

The fungal hyperparasites are also called 'mycoparasites'.

Contagious Hypovirulence

Some of the viruses which are called 'mycoviruses' attack fungi and cause reduction in their virulence. This phenomenon is called contagious hypovirulence.

MANAGEMENT OF SOIL-BORNE DISEASES BY BIOLOGICAL CONTROL

(I) Annulment of Fungistasis

Soil-borne plant pathogens survive in the soil for long time mainly because of soil fungistasis which inhibits germination of oospores, sclerotia and chlamydospores of the fungi. Germlings (germinating spores) are highly susceptible to lysis due to other microbes. Hence to induce lysis of fungal propagules, the fungistasis should be first annulled. Stimulation of germination of fungal propagules by organic materials without supporting formation of new resistant propagules may provide the basis for biological control. However very high organic amendments may be needed to annul the fungistasis.

Some stimulatory volatile substances have been found to annul soil fungistasis. Sclerotia of the onion white rot pathogen *(Sclerotium cepivorum)* do not germinate in soil because of fungistasis. Volatiles such as organic sulfides are exuded by roots of onion and they stimulate germination of sclerotia of *S. cepivorum.* Volatile organic sulfides and their precursors are now commercially available. The synthetic sulfide-precursors persist in natural soil and control the disease by inducing sclerotia to germinate in the absence of the host. The germlings die either from nutrient exhaustion or lysis by microorganisms.

Volatile compounds liberated during alfalfa hay decomposition increase microbial activity in soil, stimulate germination and lysis of sclerotia of *Sclerotium rolfsii.* Thus incorporation of alfalfa hay in soil controls soil-borne diseases.

(II) Elevation of Fungistasis

Instead of annulling fungistasis, we can elevate fungistasis of soils and control soil-borne diseases. Soil can be made increasingly fungistatic to chlamydospores of *Thielaviopsis basicola* by amending it with alfalfa hay and allowing the decomposition to proceed for more than 4 days. Rye and corn residues enhance fungistasis to chlamydospores of *Fusarium solani* f. sp. *phaseoli.* These treatments elevate the levels of toxicity of soil to such an extent that the propagules become inactivated even in the presence of root exudates.

(iii) Inhibitory Volatiles

Crucifers when added to the soil, many volatiles such as methanethiol, dimethyl sulfide, and dimethyl disulfide are released. These compounds control pea root rot caused by *Aphanomyces euteiches* by inhibiting growth, zoospore mortality and germination of the fungal propagules.

(iv) Antibiotics

Many soil-dwelling microorganisms produce antibiotics in culture. But in soil antibiotic production is negligible. Large addition of organic matter may be needed to induce antibiotic production in soil.

The antibiotic, pyrrolnitrin produced by *Pseudomonas fluorescens* obtained from the rhizosphere of cotton plants, reduces cotton root rot caused by *Rhizoctonia solani* when used as seed treatment.

(v) Antagonistic Organisms

Organic amendments with high C/N ratios increase antagonistic *Streptomyces* spp. and control root rot of bean caused by *Rhizoctonia solani.* Similar amendments may be useful in the control of some soil borne diseases.

(vi) Cross Protection

When avirulent strains of the pathogen are inoculated into soil, the disease caused by virulent strains is reduced. For example, *Verticillium* wilt of cotton is controlled by incorporating avirulent strain of the pathogen in the soil.

(vii) Bacteriocin

Bacteriocins are non-replicating, bactericidal protein-containing substances which are produced by certain strains of bacteria and are active against some other strains of the same or closely related species. *Agrobacterium tumefaciens* causes crown gall in many crops. *Agrobacterium radiobacter* is a non-pathogenic bacterium and it produces bacteriocin. Bacteriocins produced by *Agrobacterium* spp. are called agrocins. When seeds or roots of peaches are inoculated with *A. radiobacter,* crown gall of peaches is effectively controlled. Similarly when non-pathogenic *A. tumefaciens* strain 84 was inoculated on to tomato the crown gall disease caused by virulent strain of *A. tumefaciens* is controlled.

IMPORTANT BIOCONTROL MICROBIAL AGENTS

1. Trichoderma Spp

(a) **Factors affecting their bioefficacy:** *Trichoderma* spp. are widely distributed all over the world and occur in all soils. *T. harzianum* and *T. viride* are the two important *Trichoderma* species which have been developed as biocontrol agents. They are found on well-decomposed organic matter, root surfaces of various plants, on decaying bark and on sclerotia of other fungi. Conidia of *Trichoderma* do not survive well in soil but chlamydospores survive for a long time.

Fungistasis suppresses the growth of *Trichoderma* in soil. Hence if *Trichoderma* were to be utilized for biological control, fungistasis in soil should be annulled. The addition of a food base along with the fungus may allow spores to germinate, but subsequent growth may not occur. Organic amendments may annul fungistasis. However, they do not support growth of *Trichoderma* because these organic amendments support other saprophytes which suppress *Trichoderma*. *Trichoderma* cannot compete with other saprophytes and it is only a secondary invader. Mature compost supports *Trichoderma* well.

Trichoderma proliferates when added as young mycelia in intimate contact with a food base (sterile moist bran inoculated with contact conidia and allowed to incubate for one to three days before addition to soil), but not as conidia with or without bran. Young hyphae are less sensitive to fungistasis. The thorough colonization of the food base by the mycelia enables *Trichoderma* to grow well through the soil. Thorough primary colonization of a substrate by *Trichoderma* before addition to soil is necessary.

Alginate pellets containing *Trichoderma* established in the bran medium also stimulate the establishment of *Trichoderma* in soil. If soil is fumigated with carbon disulfide or formaldehyde or methyl bromide *Trichoderma* multiplies well in the soil. Sodium azide treatment also increases multiplication of *Trichoderma* in soil. These chemicals are non-inhibitory *to Trichoderma* but inhibit other saprophytes.

Treatment of soil with steam enhances *Trichoderma* in soil. Covering the soil with polythene sheets enhances soil temperature (solarization) and this sub-lethal heating of soil encourages *Trichoderma*.

Thiram, captan, PCNB, chlorothalonil and chloroneb are not inhibitory to *Trichoderma* but suppress other saprophytes. However, benomyl, carbendazim, thiophanate methyl and thiabendazole are inhibitory to *Trichoderma*.

Trichoderma does not survive in the rhizosphere. *Pseudomonas* spp. compete with *Trichoderma* for iron in the rhizosphere and for want of iron the fungus does not survive in the rhizosphere. *Trichoderma* is aerobic and requires oxygen for its growth. The fungus does not survive in anaerobic soils (paddy fields).

(b) **Method of application:** Wheat bran plus peat, barley grain and composted hardwood bark are good substrates for production of *Trichoderma*. Alginate pellets with bran are useful for introduction of *Trichoderma* into the soil. Diatomaceous earth granules impregnated with molasses also serve as carrier of *Trichoderma* into the soil.

Polyethylene sheet mulching or treatment of soil with heat encourages *Trichoderma* development.

The fungus survives well in spermosphere (region around germinating seed). Hence seed treatment with the fungus controls many seed rots and damping-off.

(c) **Control of soil-borne diseases by Trichoderma:** Seed treatment with *Trichoderma* controls damping off of peas, tomato and tobacco caused by *Pythium* spp. and *Rhizoctonia solani*.

Application of *Trichoderma* in the field controls root rot *(Sclerotium rolfsi)* of groundnut.

(d) **Mode of action of Trichoderma:** *Trichoderma* produces three antibiotics, viz. trichodermin, gliotoxin and viridin in culture. But they are not produced in appreciable quantities in soil.

Trichoderma invades hyphae of *Pythium* and *Rhizoctonia* and produces haustoria. It produces β-(1-3)-glucanase and chitinase and degrades the glucans in the cell walls of *Pythium* spp. and the chitin and glucans in the walls of *Rhizoctonia*. Thus these enzymes lyse the hyphae of the pathogens.

(e) Limitations in the Use of Trichoderma as Biocontrol Agents:

(i) *Trichoderma* does not survive in rhizosphere as well as in unamended soil. Organic amendments are also not helpful. Enough bran or molasses have to be added in the field for successful biological control. The fungal inoculum also should be in young mycelial stage and old cultures with spores will be inactive. Hence storage of active cultures is difficult. Steaming or fumigating the soil to encourage *Trichoderma* will be expensive. Polythene mulching will also be expensive.

(ii) *Trichoderma* doesn't survive in anaerobic soil. Hence, it is not useful against rice diseases.

(iii) As seed treatment *Trichoderma* effectively controls soilborne diseases occurring in early stages of crop growth. However, many seed treatment fungicides inhibit *Trichoderma.*

2. Other Fungi

Gliocladium virens acts similar to *Trichoderma* as biocontrol agent against *Pythium, Sclerotium* and *Rhizoctonia* diseases. *Sporidesmium sclerotivorum* controls *Sclerotinia* diseases. It controls lettuce drop in the field caused by *Sclerotinia minor.*

Coniothyrium minutans controls *Sclerotinia sclerotium* in soil.

3. Bacteria

Siderophores: Many bacteria produce siderophores. Siderophores are low molecular-weight ferric iron-transport agents. The function of siderophores is to supply iron to the bacterial cell. Siderophores selectively complex with iron with very high affinity and make it unavailable to other microorganisms including pathogens. *Pseudomonas fluorescens* and *P. putida* produce siderophores. These two bacteria control soft rot of potato caused by *Erwinia carotovora.* The siderophores of *Pseudomonas efficiently* sequester ferric iron in potato root zone making it unavailable to other rhizoplane microorgnisms, including plant pathogens. Hence the populations of the pathogen are reduced and the roots appear healthier.

The siderophore of *Pseudomonas* sp. is fluorescent and it is called *Pseudobactin. Pseudobactin* has been found to control take-all disease of wheat and barley caused by *Gaeumannomyces* var. *tritici* and flax wilt disease caused by *Fusarium oxysporum* f. sp. *lini.* Inoculation of flax seeds with a suspension of *Pseudomonas* sp. or pseudobactin controls *Fusarium* wilt disease. Addition of pseudobactin to the soil also controls wilt diseases.

The pseudobactin converts pathogen-conducive soils into pathogen suppressive soils. It suggests that suppressiveness of the soil is mainly due to the siderophores.

BIOLOGICAL CONTROL OF AIR-BORNE DISEASES

(a) Phyllosphere Microflora

On the leaf surface (phylloplane) many bacteria and fungi survive as saprophytes. *Erwinia, Pseudomonas, Xanthomonas, Flavobacterium, Bacillus, Lactobacillus* and *Corynebacterium* are the common bacteria found in the phylloplane. Yeasts are also commonly found. In general, fungi do not form colonies on the leaf surface. However, in the senescent leaves *Cladosporium* is commonly present. When honey dew (insect's excretion) and pollen are deposited on leaves *Cladosporium, Aureobasidium, Cryptococcus* and *Penicillium* spp. occur predominantly.

(b) Saprophytes as Scavengers of Extraneous Nutrients

Aphids' honey dew increases *Septoria nodorum* infection in wheat. But the saprophytes found on the phylloplane nullify the stimulation. Similarly pollens falling on leaves increases *Helminthosporium* leaf spot of rye. But the saprophytes scavenge the nutrients on the phylloplane and decrease the disease.

Sometimes foliar spray with fungicides increase the disease instead of decreasing it. For example, benomyl spray increases *Helminthosporium sativum* leaf spot of rye because benomyl is ineffective against *H. sativum* but eliminates many beneficial saprophytes like *Penicillium* spp.

(c) Hyperparasites

The hyperparasitic fungi attack hyphae and sporulating structures of the plant pathogens. Important hyperparasitic fungi are *Darluca filum, Tuberculina vinosa* and *Verticillium lecanii* on rusts and *Ampelomyces quisqualis* on powdery mildews. *Darluca filum* occurs naturally on uredo and teleutosori of numerous rust species. *V. lecanii* controls coffee rust *(Hemileia vastatrix)* and wheat stem rust *(Puccinia graminis tritici)* . *Ampelomyces quisqualis* controls powdery mildew of cucumber.

Hyperparasites require prolonged moist weather for their effective control of diseases. Application of urea encourages saprophytes and controls apple scab caused by *Venturia inaequalis.*

(d) Contagious Hypovirulence

Chestnut canker caused by *Endothia parasitica* is controlled by hypovirulent strains of the pathogen. The contagious hypovirulence is caused by double-stranded RNA being transmitted by virus infection to virulent strains.

(e) Predatory Bacteria

Bdellovibrio bacterivorus found on phylloplane parasitize other bacteria. In controls *Pseudomonas syringae* pv. *glycinea* infection on soybean leaves.

(f) Bacteriocin

Bacteriocins have been isolated form *Pseudomonas syringae.* They are called syringacins. When syringacin is sprayed on bean leaves prior to inoculation with *Pseudomonas syringae* pv. *phaseolicola,* the bean halo blight pathogen, the disease is controlled. Even as little as 3 ng per leaf controls the disease. Bacteriocin isolated from *Corynebacterium michiganense* controls tomato canker caused by *C. michiganense* itself.

(g) Bacteriophages

When bacteriophages are sprayed, marked protection against various bacterial diseases has been reported. However, application of phage simultaneously with the pathogen or after inoculation of the plant with the pathogen is ineffective.

(h) Introduction of Naturally-occurring Antagonists

Preinoculation of apple blossoms with *Erwinia herbicola* gives partial control of fire blight of apple caused by *E. amylovora.* *E. herbicola* produces bacteriocin-like substances which inhibits growth of *E. amylovora in vitro.*

Preinoculation of *Erwinia* and *Pseudomonas* suppresses rice bacterial leaf streak pathogen, *Xanthomonas campestris* pv. *oryzicola.* Incompatible strains of *X. campestris* pv. *oryzae* inhibit lesion enlargement on rice leaves by compatible strains of the bacterium.

(i) Introduction of Foreign Antagonists

When the soil-borne *Trichoderma viride* is applied on grapevine, it colonizes dead plant remains such as flowers which would otherwise provide a food base for *Botrytis* infection. By this method *Botrytis* rot of grapevine is controlled.

Botrytis cinerea eyespot of apples is controlled by spraying conidia of *Trichoderma pseudokoningii.* Tree wound application of *T. viride* is known to control silver leaf disease of plum caused by *Stereum purpureum.* For this, *T. viride* impregnated with wood dowels (pegs that fit into two corresponding holes to join two adjacent parts) are introduced into holes made at 10 cm intervals in the trunk.

REFERENCES

1. Blakeman, J.P. and N.J. Pokkema, 1982. Potential for biological control of plant diseases on the phylloplane. Annu. Rev. Phytopathol, **20**: 167-192
2. Bochandi, E. 1975. Biological control of bacterial canker of tomato with bacteriocins from *Corynebacterium michiganense.* Ann. Meet., Am. Phytopathol. Soc. Aug., 10-14, No. 154
3. Ersinke, J.M. and L.E. Lopateeki. 1975. *In vitro* and *in vivo* interactions between *Erwinia amylovora* and related saprophytic bacteria. Can. J. Microbiol, **21**: 35-41
4. Howell, C.R. and R.D. Stipanovic, 1979. Control of *Rhizoctonia solani* on cotton seedlings with *Pseudomonas fluorescens* and with an antibiotic produced by the bacterium. Phytopathology, **69**: 480-482
5. Htay, K. and A. Kerr, 1974. Biological control of crowngall: Seed and root inoculation. J. Appl. Bacteriol, **37**: 525-530
6. Huang, H.C. 1977. Importance of *Coniothyrium minitans* in survival of sclerotia of *Sclerotinia sclerotiorum* in wilted sunflower. Can. J. Bot., **55**: 289-295
7. Kerr, A. and K. Htay, 1974. Biological control of crown gall through bacteriocin production. Physiol. Plant Pathol, 4: 37-44
8. Lockwood, J.L., 1977. Fungistasis in soils. Biol. Rev., **52**: 1-43
9. Mew, T.W., J.S. Huang and E. Echandi, 1982. Detection of bacteriocin-like substances produced by *Xanthomonas campestris* pv. oryzae. Phytopathology, **72**: 946
10. Papavizas, G.C. 1985. *Trichoderma* and Gliocladium: Biology, ecology and potential for biocontrol. Annu. Rev. Phytopathol, **23**: 23-54
11. Papavizas, G.C. and R.D Lumsden, 1980. Biological control of soilborne fungal propagules. Annu. Rev. Phytopathol. **18**: 389-413
12. Shipton, P.J., 1977. Monoculture and soil-borne plant pathogens. Annu. Rev. Phytopathol, **15**: 387-407
13. Vidaver, A.K., 1976. Prospects for control of phytopathogenic bacteria by bacteriophages and bacteriocins. Annu. Rev. Phytopathol, **14**: 451-465
14. Vidaver, A.K., 1983. Bacteriocins: The lure and the reality. Plant Dis., **67**: 471-475
15. Watanabe, M., K. Nakanishi, K. Kumakura and N. Shimoyama, 1976. Studies on the mechanism of resistance in rice plants infected with *Xanthomonas oryzae.* Inhibition of lesion enlargement by the pre-inoculation with non-pathogenic bacteria. Ann. Phytopath. Soc. Japan, **42**: 295-303
16. Watson, A. G. And E. J. Ford, 1972. Soil Fungistasis - A reappraisal. Annu. Rev. Phytopathol, **54**: 113-144.

19

DISEASE MANAGEMENT — FUNGICIDES AND ANTIBIOTICS

DEFINITIONS

Fungicide is a chemical which is capable of killing fungi.

Antibiotic is a chemical produced by a microorganism which is inhibitory to other microorganisms. Fungicides can be classified as protectants, eradicants and systemic fungicides.

Protectant Fungicide which is effective only if applied prior to fungal infection is called protectant. *Examples* — Mancozeb, Zineb.

Eradicant is the one which removes fungi from an infection court. An eradicant can penetrate the host tissues to a limited extent and eliminate on established infection. It is also called topical therapeutant. *Examples* — Lime sulphur, organomercurials. They form vapour, penetrate into the plant cells and eradicate the fungi.

Systemic Fungicides

Systemic fungicides are the compounds which are transported over a considerable distance in plant system after-penetration. They kill fungi which are found remote from the point of application. They are chemotherapeutants. Chemotherapeutants are the compounds which eliminate a pathogen after it has penetrated into the host.

Fungicides are mostly in the form of wettable powders or dusts.

Wettable Powders

The wettable powders are water-dispersible and they remain well dispersed in water for a long time. The fungicide formulations are not prepared as water-soluble because on foliage they will be easily washed out by rain drop or dew. The wettable powder besides containing active ingredient of the fungicide, contains one wettable agent (1-10% polyethylene oxide condensate), a thickening agent like carboxymethyl cellulose and a hydrophilic diluent such as bentonite.

Dusts

The fungicides should be in fine particles to be used as dust formulations. The active ingredient of the fungicide is carried on a carrier particle. The common carriers (diluents) used are Attapulgite, kaolin, talc, pyrophyllite, diatomaceous earth and bentonite. Sulphur dust normally does not require a carrier. However, a conditioning agent such as magnesium carbonate or talc is used.

CHARACTERISTICS OF FUNGICIDES

The characteristics of a good fungicide are listed below:

1. **Fungitoxicity:** The fungicide should be fungitoxic even at low concentrations.
2. **Coverage:** The fungicide falling on the host should cover the entire host surface. The particles size of the fungicide should be very small. Small size of the particle will facilitate easy distribution on the host surface. Only when the fungicide covers the entire surface, it can protect against maximum number of potential infection loci.
3. **Redistribution:** Often the droplets will coalesce and run off. Due to this fungicide will accumulate on the tips and mid-veins of leaves rather than on the edges and interiors of the blades. The coalesced fungicide should get redistributed by dew or rain. The rain should not wash away the fungicide, but should redistribute on the entire surface of leaves.
4. **Tenacity:** The fungicide spray-deposit should adhere to the leaf surfaces. It should persist for a long time and resist the weathering process.
5. **Phytotoxicity:** The fungicide should not be phytotoxic.
6. **Mammalian toxicity:** The fungicide should have very low mammalian toxicity.
7. **Storability:** The fungicide should not get degraded quickly during storage.

Some adjuvants are added to the fungicide to improve the quality of fungicides.

Wetting Agents

The leaf surface has a waxy cuticle layer. The wax has no affinity for water. Hence, when the fungicide in water suspension is sprayed on the leaves, it fails to wet the cuticle. Leaf hairs on the cuticle layer trap pockets of air between the aqueous liquid and the cuticle. It prevents wetting. Wetting agents reduce surface tension of particles. They also prevent the formation of a layer of air between a solid and a liquid. The wetting agents when added to the fungicidal preparation help the fungicides to get deposited on leaves.

Polyethylene oxide condensate esters of fatty acids and flour are the important wetting agents.

Spreaders

Spreaders help to improve the coverage of fungicides on leaf surface. Soap and flour are good spreaders.

Stickers

Stickers increase adhesion of fungicides to host surfaces. They increase the tenacity of fungicides. Polyethylene polysulphide, gelatine, casein, fish oil and linseed oil are good stickers.

Deflocculating Agents

Deflocculating agents help the dispersion of fungicide particles in water. They keep particles away from each other and prevent solidification. They retard sedimentation of particles in the spray suspension. Gelatine and plant gums are good deflocculating agents.

Fungicides are classified based on their chemical structures.

1. SULPHUR FUNGICIDES

(a) Inorganic Sulphur Fungicides

(i) **Elemental sulphur:** Two types of sulphur formulations are available. They are sulphur dusts and wettable sulphur. The efficacy of sulphur dust depends upon the fineness of its particles. The particle size should be less than 75 microns. The sulphur particles are mixed with small amounts of inert materials such as kaolin or lead arsenate (carriers).

(ii) **Wettable sulphur:** Wettable sulphur is industrially prepared by grinding it with substances like casein, bentonite clay and sulphite lye.

(iii) **Lime sulphur:** Lime sulphur can be prepared by the farmers themselves by mixing lime and sulphur and boiling them. Commercial formulations of lime sulphur are also available.

Inorganic sulphur fungicides emit vapour and prevent spore germination and germ tube elongation. Because of vapour action, sulphur will be active even at a distance from deposits on leaves. Sulphur fungicides effectively control powdery mildews of different crops such as chilli, okra, grape, rubber, mango, citrus, blackgram and greengram. Sulphur controls tikka leaf spot of groundnut, and *Diplocarpon* black spot of rose. Wettable sulphur is sprayed at about 2.5 kg/ha. Sulphur dust is applied at about 25 kg/ha.

Sulphur dust is used as seed treatment also. When sorghum seeds are treated with sulphur dust at 5 g/kg of seed, grain smut is effectively controlled. Sulphur dust is dusted to the soil at 100 kg/ha to control powdery mildew of tobacco. Sulphur dust emits vapour and controls the disease. Sulphur is toxic to tobacco leaves and hence soil application is done. Potato scab *(Streptomyces scabies)* is controlled by applying sulphur to the furrows at 500 kg/ha. Sulphur reduces pH of the soil, which is not conducive for the development of this soil-borne pathogen.

(b) Organic Sulphur Fungicides

Dithiocarbamates are the organic sulphur fungicides. They are divided into two groups viz. dialkyldithiocarbamates (thiram, ziram and ferbam) and monoalkyldithiocarbamates (maneb, zineb, mancozeb, vapam and nabam).

Dialkyldithiocarbamates

1. **Ziram** (Zinc dimethyldithocarbamate): It is marketed in the names of Cuman, Hexazir, Ziram and Corozate. It is used as a protectant and sprayed before the outbreak of the disease. It controls early blight of potato and tomato *(Alternaria solani)* and anthracnose diseases of beans *(Colletotrichum lindemuthianum)* and cucurbits *(C. lagenarium)*. The normal dosage used is 0.2% (1 kg/ha).
2. **Thiram** [Tetramethylthiuram disulphide (TMTD)]: It is marketed in the names of Thiride, TMTD, Hexathir, Arasan and Thiram.

It is commonly used for seed treatment. It is used both as dry powder or slurry for seed treatment. The dosage used is about 2 g/kg seed. Thiram seed treatment controls seed-borne pathogens as well as soil-borne pathogens. It controls seed-borne infection of *Colletotrichum capsici* of chilli, root rot of groundnut *(Macrophomina phaseolina)*, *Alternaria raphani* of radish, sorghum grain smut *(Sphacelotheca sorghi)* and *Helminthosporium oryzae* of rice. As seed treatment it controls soil-borne infection of *Pythium* spp. of tomato, tobacco and brinjal (damping off), *Rhizoctonia solani* of cotton (root rot) and *R. solani* of rice (sheath blight).

Thiram seed soaking method is found effective in controlling internally seed-borne pathogens. The seeds are soaked in 0.2% thiram for 24 hrs. and then dried. Among the non-systemic fungicides thiram is the only fungicide which has been found to penetrate the seed coat and eradicate internal mycelia. It is effective against *Ascochyta pisi* in pea, and *Phoma lingam* and *Alternaria brassicicola* in cabbage.

Thiram is used as a foliar spray for the control of sugarbeet rust *(Uromyces betae)* at 0.5%, and *Alternaria brassicae* of cabbage at 0.2%. It is used for drenching the soil also for the control of *Pythium aphanidermatum* of red pepper and seedling diseases of cotton at about 0.2%.

Ferbam (Ferric dimethyldithiocarbamate): It is marketed in the names of Ferbam, Hexaferb and Coromet. It is used to control diseases of apple. It controls leaf spot of banana *(Deightoniella torulosa)*, leaf mould of tomato *(Cladosporium fulvum)* and leaf spot of coffee *(Cercospora coffeicola)*. It is used as a spray at about 0.2% concentration.

Monoalkyldithiocarbamates

Zineb (Zinc ethylenebisdithiocarbamate): It is marketed in the names of Dithane Z-78, Hexathane, Lonacol and Parzate C. It contains zinc and plants can absorb zinc from this compound. Hence it is beneficial to correct zinc deficiency also. It is commonly used as a foliar spray at 0.2%. It controls anthracnose diseases of bean, chilli and cucurbits, rust diseases of wheat, sorghum and bajra, downy mildews of grapevine, cucurbits, onion and cabbage, *Cercospora* leaf spots of groundnut, cabbage, cauliflower, *Alternaria* leaf spots of potato, tomato and ginelly and rice blast and brown leaf spot. Zineb is not stable for long time and hence it has to be sprayed once in 7-10 days.

Mancozeb and Maneb (Manganous-zinc ethylene bisdithiocarbamate, manganous ethylene bisdithiocarbamate): Mancozeb is marketed in the name of Dithane M-45. Maneb is marketed in the names of Dithane M-22 and Manzate. Another mixture of maneb (53%) with nickel sulphate (19%) is marketed in the name of Dithane S-31.

Mancozeb is widely used for the control of late blight of potato *(Phytophthora infestans)*. It is a versatile fungicide, used against almost all kinds of diseases. In particular, it is effective against *Cercospora* leaf spots of groundnut, cucurbits and sugarbeet, *Helminthosporium* leaf spots of rice, ragi, maize and sorghum, downy mildews of grapevine and tobacco, rust diseases of wheat, groundnut, bajra and sorghum, *Alternaria* leaf spots of gingelly, potato, tomato and wheat, anthracnoses of chilli, grapevine, sorghum, bean and cucurbits and *Pyricularia* (blast) of rice and ragi. It controls white rust of amaranthus and crucifers, *Septoria* leaf spots of tomato and wheat and black spot of rose *(Diplocarpon rosae)*.

Mancozeb is used as seed treatment also. As seed treatment at 2g/kg seed, it controls stackburn disease of rice *(Alternaria padwickii)* and bunts of wheat *(Tilletia caries* and *T. foetida)*.

Nabam (Disodium ethylene bisdithiocarbamate): It is commerically marketed as Dithane D-14 and Parzate liquid. It is highly soluble and unstable. Hence nabam is not being used widely at present.

Vapam (Sodium methyldithiocarbamate): It is highly soluble in water and cannot be used as foliar spray. It is used to drench the soil to control soil-borne pathogens. When it is applied to the soil, it releases the fumigant, methylisothiocyanate. It should be drenched into the soil before growing the crops. It controls cotton wilt and damping off of papaya, tobacco and tomato. It controls nematodes also.

2. COPPER FUNGICIDES

(a) Copper Sulphate Preparations

Bordeaux mixture: In 1882, Dr. Millardet discovered the use of Bordeaux mixture in the control of downy mildew of grapevine *(Plasmopara viticola)* in France. Since then it is the most widely used fungicide in the control of several diseases in the world. Bordeaux mixture is prepared as follows: One kg of copper sulphate is powdered and one kg of lime is slaked separately. One kg of copper sulphate is added to 50 litres of water while 1 kg of lime is added to another 50 litres of water. Copper sulphate solution is then added to lime water. The prepared 1% Bordeaux mixture should be neutral. Free copper is phytotoxic and the presence of free copper can be tested by dipping an iron sickle in the mixture for some time. If copper deposits are found on the sickle, more lime should be added.

Bordeaux mixture should be prepared fresh every time before spraying. It should not be prepared in iron or tin containers. It can be prepared in plastic buckets. Wooden or earthen vessels can also be used. Addition of jaggery or sugar at 100 g/100 litre of mixture will prevent quick deterioration of Bordeaux mixture.

Bordeaux mixture 1% is highly effective against late blight of potato, downy mildew of grapevine, coffee rust *(Hemileia vastatrix)*, betelvine wilt *(Phytophthora parasitica)*, pepper wilt *(Phytophthora palmivora)*, tomato early *(Alternaria solani)* and late blights *(Phytophthora infestans)* and coconut wilt *(Ganoderma lucidum)* and bud rot *(Phytophthora palmivora)*. Bordeaux mixture 1% is sprayed on foliage to control leaf spot diseases and drenched in soil to control wilt diseases. It is useful to control bacterial diseases also. Fire blight of apple *(Erwinia amylovora)* and citrus canker *(Xanthomonas citri)* are controlled by Bordeaux mixture.

Bordeaux paste

Bordeaux paste is nothing but 10% Bordeaux mixture. It will be in a paste form. It is applied as wound dressing in trees to prevent fungal attacks.

Burgundy mixture

Instead of lime in Bordeaux mixture, sodium carbonate is used for the preparation of Burgundy mixture. However it is less efficaceous than Bordeaux mixture.

Cheshnut compound

It is prepared by mixing 2 parts of copper sulphate with 11 parts of ammonium carbonate. Three grams of this mixture is dissolved in a little hot water and diluted to 1 litre with cold water. It is used to drench the soil to control damping-off diseases.

(b) Copper Carbonate Preparations

Chaubattia paste: Copper carbonate (800 g) is mixed with red lead (800 g) in 1 litre of lanolin or linseed oil. A paste is formed. It is used to cover pruned parts of apple, pear, peach, plums and apricot. It controls many fungal diseases of these trees. The paste is not washed away by rain.

(c) Copper Oxychloride Preparations

Many commercial formulations of copper oxychloride are available in the market. They are Fytolan, Blue copper, Blitox, Cupramar and Micop. Copper oxychloride is oxidized to copper hydroxide when copper oxychloride solutions are exposed to air. Copper oxychloride fungicides are low soluble fingicides (less soluble in water). They are generally used at 0.25% concentration and effective against all diseases against which Bordeaux mixture has been found effective. However efficacy is lower than Bordeaux mixture because of low tenacity.

(d) Cuprous Oxide Preparations

They are marketed as Fungimar and Perenox. They are also low-soluble fungicides. They are used at 0.3% concentration as spray.

3. MERCURY FUNGICIDES

Organic mercurial compounds are generally used as fungicides. Ethyl mercury chloride is sold in the market as ceresan, ethyl mercury phosphate as New Improved ceresan, methoxyethyl mercury chloride as Aretan, Agallol and Wet ceresan, phenyl mercury acetate as Dry ceresan, hydroxy mercurichlorophenol as Semesan, and methylmercuric dicyandiamide as Panogen. All these fungicides are effective as seed treatment. They have vapour action and hence eliminate even deep-seated infection.

Ceresan and Agrosan dry seed-treatment effectively control *Helminthosporium* seedling blight of rice, wheat, ragi and oats, *Aspergilus niger* seedling blight of groundnut, sorghum smut *(Sphacelotheca sorghi)*, bunts of wheat *(Tilletia carries* and *T. foetida)* and loose smut of oats *(Ustilago avenae)*. Wettable formulations of organic mercurials like Aretan or Agallol (6% mercury) are used to treat sugarcane setts to control pine apple disease *(Ceratocystis paradoxa)*. Wet ceresan is used as soil drench to control soil-borne pathogens like *Rhizoctonia solani, Macrophomina phaseolina, Fusarium* spp. and *Pythium* spp. which affect several crops.

Mercurials are phytotoxic and hence not used as foliar sprays or dusts. However a specific formulation of phenyl mercury acetate dust has been found to control rice blast and apple scab *(Venturia inaequalis)*.

4. HETEROCYCLIC NITROGEN COMPOUNDS

(i) **Captan (N-trichloromethyl thio-4-cyclohexene-1, 2-dicarboximide):** It is commercially marketed as Captan, Orthocide, Vancide and Esso fungicide 406. It is a protectant fungicide and persists up to 10 days. It is degraded when exposed to sunlight. Captan controls maize helminthosporiose *(Helminthosporium turcicum)*, chilli fruit rot *(Phytophthora capsici)* and apple scab *(Venturia inaequalis)* when sprayed at 0.2%. It is mostly used as seed treatment. As dry seed dressing at 2 g/kg seed it controls groundnut root rot *(Rhizoctonia bataticola)* and seedling blight *(Aspergillus niger)* and damping off of beans, cucurbits, bengalgram and chilli *(Pythium* and *Phytophthora* spp.).

(ii) Folpet [N - (trichloro-methyl-thio)] phthalimide: It is commercially marketed as Phaltan. It controls rose black spot *(Diplocarpon rosae)* and apple scab *(Venturia inaequalis)* as foliar spray at 0.2%.

(iii) *Captafol* [cis-N-1,1,2,2-tetrachloroethylthio) 4-cyclohexane 1,2-dicarboximide]: It is marketed as Difolatan, Foltaf, etc. It is mostly used as foliar sprays. Captafol persists for a long period on foliage. It is sprayed at about 0.15%. It effectively controls sheath rot of rice *(Sarocladium oryzae)*, grain discolouration of rice *(Helminthosporium oryzae* and *Curvularia lunata)* and mango anthracnose *(Colletotrichum mangiferae)*.

5. QUINONE FUNGICIDES

(i) Chloranil (2, 3, 5, 6-tetrachloro-1,4-benzoquinone): It is commercially marketed as Spergon. It is a good seed dressing fungicide and not used as foliar spray. It controls grain smut of sorghum *(Sphacelotheca sorghi)* and damping off of beans *(Pythium* sp.) and cotton *(Rhizoctonia solani)*. It is used to control some soil-borne diseases by drenching the soil at 0.1% concentration.

Cinchona damping-off caused by *Rhizoctonia solani* is effectively controlled.

(ii) *Dichlone* (2,3-dichloro-1, 4-napthoquinone): The commercial name of the fungicide is Phygon. It is also used as seed protectant. Sorghum grain smut is controlled by seed treatment with it at 1.2 g/kg seed. It controls peach leaf curl *(Taphrina deformans),* apple scab *(Venturia inaequalis)* and bean anthracnose *(Colletotrichum lindemuthianum)* when sprayed at 0.2%. It is also used at 0.5% concentration for treating sugarcane setts to control pineapple disease *(Ceratocystis paradoxa).*

6. MISCELLANEOUS FUNGICIDES

(i) **Quintozene** (Pentachloronitrobenzene): Commercial names of the fungicide are Brassicol, Terraclor, PCNB and Tritisan. It is used to control soil-borne pathogens when applied as soil drench at 0.25 to 0.3%. It is effective against *Rhizoctonia solani, Macrophomina phaseolina* and *Sclerotinia sclerotiorum* and ineffective against *Pythium, Phytophthora* and *Fusarium* spp.

(ii) **Dinocap** (2, 4-dinitro-6-octylphenyl crotonate + 2, 6-dinitro 4-octylphenyl crotonate): It is marketed in the name of Karathane, Arathane, Mildex, Capryl, etc. It is generally available in liquid formulation. Dinocap is effective in controlling powdery mildews at 0.05 to 0.1% spray concentration. Powdery mildews of chilli, apple, cucurbits, crucifers, okra, grapes, blackgram, greengram, rose, tobacco and wheat.

(iii) **Chinomethionat** (6-methyl-quinoxaline-2, 3-dithol-cyclic carbonate): It is sold as Morestan in the market. It controls powdery mildews effectively when sprayed at about 0.05%. Powdery mildews of rose, tobacco, cucurbits, mustard and peas have been reported to be controlled effectively by this fungicide.

(iv) **Fenaminsosulph** (Sodium p-dimethylaminobenzene-diazo sulphonate): It is commercially known as Dexon. It is highly effective against Phycomycetes like *Pythium, Phytophthora* and *Aphanomyces* when applied as soil drench at 0.01%.

(v) **Edifenphos** (O-ethyl-S S-diphenyl-dithiophosphate): It is marked as Hinosan. It is highly effective against *Pyricularia oryzae* and it controls rice blast when sprayed at 0.1%. It also controls rice brown spot *(Helminthosporium oryzae).*

(vi) **Fentin hydroxide** (triphenyl tin hydroxide): The commercial name of the fungicide is Duter. It effectively controls tikka leaf spot of groundnut *(Cercospora personata* and *C. arachidicola)* and rust of groundnut *(Puccinia arachidis),* potato early *(Alternaria solani)* and late *(Phytophthora infestans)* blights and rice blast *(Pyricularia oryzae)* when sprayed at 0.2% concentration.

(vii) **Triphenyltin acetate:** It is marketed in the name of Brestan. It controls brown spot of rice *(Helminthosporium oryzae),* groundnut tikka leaf spot *(Cercospora personata* and *C. arachidicola)* and rust *(Puccinia arachidis)* and *Septoria* leaf spot of celery when sprayed at 0.1%. It controls *Phytophthora, Pyricularia* and *Alternaria* diseases also.

(viii) **Dicloran** (2,6-dichloro-4-nitroaniline): Its trade name is Botran. It controls *Botrytis* infection in several crops effectively when sprayed at 0.1%. It is effective against *Sclerotinia sclerotiorum* also.

(ix) **Dodine** (n-dodecyl guanidine acetate): The commercial name of the fungicide is Cyprex. It controls apple scab *(Venturia inaequalis)* when sprayed at 0.2%.

(x) **Chlorothalonil** (Tetrachloroisophthalonitrile): It is marketed as Daconil and Kavach. It controls both tikka leaf spot *(Cercosporidium personata* and *Carcospora arachidicola)* and rust disease *(Puccinia arachidis)* of groundnut and betelvine wilt *(Phytophthora parasitica).*

SYSTEMIC FUNGICIDES

Movement Through Leaves

Systemic fungicides enter into the host tissues and control fungal infection. They enter through leaves, roots and seed surfaces. Leaf surfaces have cuticle which is made up of hydroxy fatty acids and waxy layer. Most of the fungicides are insoluble in water but they are lipophilic. The cuticle has apolar lipophilic properties that inhibit penetration of aqueous solution but allow the pénetration of lipophilic compounds. Hence the systemic fungicides which are lipphilic cross the cuticle easily and enter into the vascular stream. Systemic fungicides do not enter through stoma. Even pure water does not penetrate stomata. However, some of the systemic fungicides like triademefon enter through stomata in the form of gases. Once the systemic fungicides enter into the vascular stream, they move passively along with the transpiration stream to leaves of various branches. They reach the regions of evaporation via cell wall of the leaf and accumulate in the tip and margin of the leaf. The plant organs which transpire more, receive more systemic fungicides and leaves transpire more than any other organ. Hence systemic fungicides accumulate more in leaves.

Upward translocation occurs through xylem vessels and most of the systemic fungicides move, only towards upwards (apoplastic movement). Downward translocation occurs through phloem vessels and the phloem translocation requires energy (it is not passive, but active). Sugars (photosynthates) are translocated downwards. Systemic fungicides should form complexes with sugar for phloem transport. Sugars move due to osmotic pressure and reach the various parts of the plant wherever they are deficient. This type of movement is called *symplastic* movement. Symplastic movement of systemic fungicides has also been observed but the movement is very much limited. Systemic fungicides sprayed on leaves move from one leaf to another above or below the sprayed leaf. But the fungicides sprayed on leaves do not move towards root and they do not accumulate in the roots or hypocotyls to prevent soil-borne infection. Very few exceptions have been reported. Pyroxychlor, when sprayed on foliage controls root diseases of tobacco (22).

Movement Through Roots

Systemic fungicides move faster through roots than through leaves. The fungicides penetrate the root tip epidermis due to their lipophilic action and enter the vascular tissues. They are carried through the transpiration stream and reach the foliage which have stomata. The fungicides do not reach flowers, fruits and seeds which do not have stomata. Systemic fungicides accumulate more in foliage, less in stem, in trace amounts in fruits and almost nil in seeds. However, in bean pods which have stomata, systemic fungicides accumulate and they are detected even in seed coat and cotyledons.

Movement Through Seeds

Seed treatment with systemic fungicides has resulted in control of seed-borne, soil-borne and air-borne pathogens. The systemic fungicides enter into the seed at the time of germination, accumulate in cotyledons and are translocated into the foliage. Translocation ability depends upon the type of fungicide. For example, seed treatment of soybean with chloroneb results in the accumulation of it in hypocotyl and cotyledons while benomyl moves to the epicotyl region and carboxin is distributed throughout the seedling. Due to quick distribution of carboxin throughout the plant the internally seed-borne smut pathogen of wheat *(Ustilago tritici)* is effectively controlled by seed treatment. Benomyl as it accumulates in hypocotyl and epicotyl region, controls *Rhizoctonia solani* and *Macrophomina phaseolina* (soil-borne pathogens) in several crops. Pyroquilon as seed treatment controls air-borne infection of *Pyricularia oryzae*, the rice blast pathogen even after 90 days of sowing.

Systemic fungicides enter into the seed only when the seeds are soaked in water for some time. Dry or slurry seed treatment will not control internally seed-borne pathogens when the seeds are in storage. But when the seeds are sown in the field, the fungicides enter into the seed probably through seedcoat, but most possibly through the splits formed during germination process. The organic solvents like dichloromethane, acetone, carbontetrachloride, benzene and ethanol have been used as carriers to infuse the systemic fungicides into the seed at dry conditions (19, 20).

Movement through Soil

When systemic fungicides are drenched in soil, they are taken up by the roots and become systemic in the plant system. Free water in the soil is necessary for the uptake of fungicides. Soil type determines the uptake of fungicides. They can move quickly in sandy soil rather than in loamy soil. Addition of surfactants like tween 20 increases the usefulness of the systemic fungicides as soil treatment. But in general the efficacy of systemic fungicides is reduced when used as soil drench. The fungicides are adsorbed in soil colloids. Soil organic matter, cation exchange capacity and soil moisture determine the efficacy of these fungicides.

Some of the systemic fungicides are listed below:

(i) **Carbendazim** (Methyl 1-2-benzimidazole carbamate, MBC)

It is marketed as Bavistin, Derosal, B-sten, etc. Carbendazim penetrates and moves through the xylem tissue. It moves upwards from the point of application. When lower leaves are sprayed, the fungicide moves to the upper leaves and not vice versa. When it is applied as root rip or soil drench or seed treatment, it is translocated throughout the plant system. However root dipping is difficult and can be done only at the time of transplanting. In soil it gets fixed and 80 to 95% applied chemical is not available to the plants and as seed treatment only limited quantity of fungicides can be fed to the plants and it gets diluted when the plant mass increases.

Carbendazim is effective against *Cercospora, Colletotrichum, Macrophomina, Rhizoctonia,* and *Fusarium* diseases effectively. It controls powdery mildews, smut diseases and bunts. It controls rice blast *(Pyricularia oryzae).* Carbendazim is not effective against *Helminthosporium, Alternaria* and *Curvularia* diseases. It is ineffective against rusts and all oomycetes and zygomycetes. *Pythium* and *Phytophthora* are not at all controlled.

(ii) **Benomyl** [methyl-1-(butyl carbamoyl)-2-benzimidazole carbamate]

It is similar to carbendazim. It decomposes slowly into MBC (carbendazim) in storage. On absorption benomyl is converted into MBC in the plant system. MBC appears to be the active principle in the plant system. Hence benomyl acts almost like carbendazim in almost all aspects.

Benomyl is effective against *Fusarium, Rhizoctonia, Macrophomina, Cercospora, Colletotrichum, Pyricularia, Verticillium, Thielaviopsis, Ceratocystis, Phomopsis, Septoria, Rhynchosporium, Erysiphe, Plasmodiophora, Botrytis, Ustilago, Urocystis* and *Tilletia* spp. It is ineffective against *Pythium, Phytophthora, Alternaria, Helminthosoprium, Sclerotium, Aphanomyces, Puccinia, Stemphylium* and *Curvularia.*

Benomyl has nematicidal properties also. It controls *Heterodera avenae, Pratylenchus, Ditylenchus, Aphelenchoides,* and *Xiphinema.* It is ineffective against *Meloidogyne.*

(iii) **Thiabendazole** [2-(4-thiazolyl) benzimidazole, TBZ]

It is commercially available as Tecto and Mertect. It controls wheat bunt *(Tilletia controversa)* when applied as seed treatment. As foliar spray it effectively controls apple scab *(Venturia inaequalis)*. It controls powdery mildews, *Diplocarpon rosae* (black spot of rose) and *Cercospora* leaf spots. It is effective against *Thielaviopsis basicola*.

(iv) **Carboxin** (5,6-dihydro-2-methyl-1, 4-oxathiin-3-carboxanlido)

It is commercially available as Vitavax. It is the most effective fungicide to control internally seed-borne loose smuts of cereals. Loose smut of wheat *(Ustilago tritici)*, covered smut of oats *(Ustilago kolleri)*, grain smut of sorghum (*Sphacelotheca sorghi)*, bunts of wheat *(Tilletia caries* and *T. foetida)*, and flag smut of wheat *(Urocystis agropyri)* are all controlled by seed treatment.

Carboxin is effective against *Helminthosporium oryzae* (rice brown spot pathogen) and *H. maydis* (maize leaf blight) as seed treatment. It also controls *Rhizoctonia solani* and *Alternaria* diseases. Carboxin is effective only as seed treatment. Foliar and soil applications are either less effective or ineffective.

(v) **Oxycarboxin** (5-6-dihydro-2-methyl-1, 4-oxathiin-3-carboxianilid-4, 4-dioxide)

It is commercially marketed as Plantvax. It is specifically effective against rust pathogens. Seed treatment with oxycarboxin protects the crops against rust diseases up to about 50 days of sowing. However rust diseases normally occur at a later stage of the crops. Hence seed treatment will be of no use. Oxycarboxin at 0.1 to 0.2% concentration effectively controls wheat rusts *(Puccinia graminis tritici, P. striiformis* and *P. recondita)*, groundnut rust *(Puccinia arachidis)*, bean rust *(Uromyces appendiculatus)* and sunflower rust *(Puccinia helianthi)*. Tea blister blight *(Exobasidium vexans)* is effectively controlled.

Although less effective, it controls smut diseases also as seed treatment.

(vi) **Pyracarbolid** (2-methyl-5, 6-dihydro-4H-pyran-3-carboxylic acid anilide)

It is commercially available as Sicarol. It is effective against rusts, smuts and *Rhizoctonia* spp. and is used as both as a seed treatment and foliar spray.

(vii) **Metalaxyl** [methyl-DL-N-(2, 6-dimethylphenyl-N-)2 methoxyacetyl - alaninate]

It is marked as Ridomil and Apron. Different metalaxyl formulations for foliar application, seed treatment and soil drenching have been developed. Metalaxyl is highly effective against phycomycetous (oomycetes) fungi like *Phytophthora, Pythium, Sclerospora, Pseudoperonospora, Plasmopara, Sclerophthora* and *Albugo*. These fungi cause damping off, blight, downy mildews and white rust. As seed treatment (1-5 g/kg seed) it effectively controls sorghum, bajra and maize downy mildews and damping off diseases of cotton, peas, tomato, etc.

Foliar application of metalaxyl controls downy mildews of sorghum, bajra, maize and grapevine. It controls late blight of potato *(Phytophthora infestans)* and white rust of radish *(Albugo candida)*. Soil application controls *Phytophthora* root rots of citrus and avocados.

(viii) **Tridemorph** (N-tridecyl-2, 6-dimethyl-morpholine)

Its trade name is Calixin. It is mainly used against powdery mildews as foliar sprays. It controls powdery mildew *(Erysiphe graminis)* of wheat, barley, and cucurbits *(E. cichoracearum)*. It controls blister blight of tea *(Exobasidium vexans)*. There is a report that it controls groundnut rust also.

(ix) **Pyroquilon** [1, 2, 5, 6-tetrahydro-4-pyrrolo (3, 2-1-i, j) quinoline 4-one].

It is commercially available as Fongorene. It controls effectively rice blast *(Pyricularia oryzae)* both as seed treatment and foliar spray. When the seeds are treated at 2 g/kg seeds, it protects the crop up to 45 days of sowing. Prolonged protection upto 90 days of sowing has also been reported.

(x) **Kitazin** (S-benzyl-oo-di-isopropyl phosphorothioate, IBP)
It is marketed as Kitazin. It is highly specific against rice blast. It is available as granular form also. Soil application of granules at 30-45 kg/ha effectively controls rice blast. As foliar application 0.05 to 0.1% is effective.

(xi) **Tricyclazole** [5-methyl-1, 2, 4-triazole (3, 4b) - benzothiazole]
It is commercially marketed as Beam. It is highly specific for the control of rice blast. It doesn't control any other disease in rice. It does not have much fungitoxicity against the blast pathogen, *P. oryzae.* But *in vivo* it is highly inhibitory to the fungus. It activates the synthesis of phytoalexins in rice plants and induces resistance against the disease. Similar to pyroquilon, it controls rice blast when applied either as seed treatment or as foliar spray. It is the most effective chemical in the control of rice blast.

(xii) **Probenazole** (3-allyloxyl-1, 2-benzoisothiazol-1, 1-dioxide)
Its commercial name is Oryzemate. It is also specific against rice blast. It also does not have fungitoxicity against the pathogen. But it induces synthesis of antifungal compounds in rice plants. Phytoalexins are synthesized due to probenazole treatment.

(xiii) **Isoprothiolane** (diisopropyl-1, 3-dithiolan-2-ylidenemalonate)
It is marketed as Fuji-1. It is also specific against rice blast. It controls rice blast either when sprayed at 500 g/ha or when applied as granules at 1-5 kg/ha. It also controls stem rot of rice *(Sclerotium oryzae).*

(xiv) **Triadimefon** [1- (4-chlorophenoxy)-3, 3-dimethyl-1-1H-(1, 2, 4-triazole-1-yl) butan-2-one]
It is commercially marketed as Bayleton. It is highly effective against powdery mildews. It also controls rust diseases. As seed treatment it controls loose smut of wheat.

(xv) **Biloxazole** [p-1-1, 1'-biphenyl-4 yloxyl-a (l, 1-dimethylethyl - 1H- 1, 2, 4-triazole-1-ethanol)]
It is marketed as Baycor. It is effective against *Cercospora* diseases and rusts. It controls sigatoka leaf spot of banana *(Cercospora musarum),* tikka leaf spot of groundnut *(Cercospora personata* and *C. arachidicola),* rust diseases of groundnut *(Puccinia arachidis)* and wheat *(Puccinia graminis tritici).* It is also effective against powdery mildews of wheat *(Erysiphe graminis),* pea *(E. pisi),* cucurbits *(E. cichoracerarum),* rose *(Sphaerotheca pannosa)* and mango *(Oidium mangiferae).*

(xvi) **Triademenol** [1-4-chlorophenoxyl-1, 3-dimethyl-1-(1, 2, 4-triazole-1-yl)- butan-2-ol]
The commercial name of the fungicide is Bayton. It is mainly used as seed treatment. As seed treatment it controls smuts and powdery mildews. It controls loose smut and bunts of wheat and powdery mildew of barley *(Erysiphe graminis).*

(xvii) **Thiophanate and Thiophanate methyl** [1,2-bis (3-ethoxycarbonyl-2-thioureido) benzene; 1,2-bis (3-methoxycarbonyl-2-thioureido) benzene]
The commercial names of thiophanate are Topsin and Cercobin while those of thiophanate methyl are Topsin M and Mildothane. These compounds are transformed into methyl-2-benzimidazole carbamate (MBC) and ethyl-2-benzimidazole carbamate (EBC) respectively. MBC is the same active principle of benomyl and carbendazim. Hence it acts similar to those fungicides. Thiopha-

nate methyl controls powdery mildew of barley *(E. graminis)*, rice blast, and stem rot of rice *(Sclerotium oryzae)*, as foliar spray. As seed treatment, it controls loose smut of maize *(Ustilago maydis)* and wheat *(Ustilago tritici)*. It controls apple scab *(Venturia inaequalis)* by foliar spray. Thiophanates are ineffective against *Helminthosporium* spp. and phycomycetous fungi.

(xviii) **Ethrimol** (5-butyl-2-ethylamino-4-hydroxy-6-methyl pyrimidin)

Its trade names are Milcurb, Milstem and Milgo. It is specifically effective against powdery mildews. It is used as seed treatment. It gives complete control of powdery mildews even for more than 15 weeks of sowing. The effective dosage is 3 to 5 g/mg of seed. As foliar sprays, it is not effective.

(xix) **Aluminium-tris** (Aluminium tris aluminium ethylphosphite)

The commercial name of the fungicide is Aliette. It does not have any fungitoxicity. But it induces synthesis of phytoalexins in plants. It is specific against Oomycetes. It is mostly effective against *Phytophthora* spp. and downy mildews. It is effective as soil drench. Interestingly it controls *Phytophthora cinnamoni* root rot of avocado when the fungicide is sprayed on foliage. It suggests downward translocation, a rare phenomenon in systemic fungicides. It also controls root rot and stem canker of citrus caused by *Phytophthora nicotianae* var. *parasitica* by foliar application of the fungicide.

(xx) **Chloroneb** (1,4-dichloro-2, 5-dimethoxy benzene)

It is marketed as Demosan. It is mostly used as seed treatment (0.5 g/kg seed) to control root rot of cotton *(Rhizoctonia solani)*. As soil treatment (15 kg/ha), it controls *Sclerotium rolfsii* root rot of sugarbeet.

(xxi) **Ethazole** (5-ethoxy-3-trichloromethyl-1, 2, 4-thiadiazole, ETMT)

It is commercially marketed as Terrazole, Truban and Terrachlor. It is effective against Oomycetes fungi such as *Phytophthora* and *Pythium*. It is also effective against *Rhizoctonia, Fusarium, Thielaviopsis*. It is used as seed treatment and soil drench.

(xxii) **Pyroxychlor** [2-chloro-6-methyl-4-(trichloromethyl) pyridine]

It is marketed as Dowco-269. It is the only effective systemic fungicide which moves downwards. It is effective against Pythiaceous fungi. Foliar application of this fungicide controls root infection of *Pythium* and *Phytophthora*.

Mode of Action of Systemic Fungicides

Systemic fungicides are selective in the control of diseases. Metalaxyl is specific against Oomycetes fungi, ethirimol against powdery mildews, carboxin against smuts, oxycarboxin against rusts, pyroquilon, tricyclazole, kitazin, probenazole and isoprothiolane against blasts and tridemorph against powdery mildews. These fungicides act on specific sites in fungal cells. Carboxin affects mitochondrial activity. Benomyl, carbendazim and metalaxyl inhibit nucleic acid synthesis. Triadimenol, biloxazol, triadimefon and tridemorph inhibit sterol synthesis in fungi. Kitazin inhibits chitin synthesis.

Such type of specificity leads to the development of fungicide-resistant strains. Benomyl-resistant tikka leaf spot pathogens *(Cercospora personata* and *C. arachidicola)* are commonly formed in fields where the fungicide is repeatedly used. To avoid this a mixture of systemic fungicides and protectant fungicides can be used. For example, a mixture of carbendazim + mancozeb can be used to control tikka leaf spot of groundnut. Systemic fungicides can be used by rotation also. For example, to control powdery mildews, benomyl is sprayed first followed by triadimenol as a second spray. Continuous use of a single systemic fungicide should be discouraged.

ANTIBIOTICS

(I) Aureofungin Sol

It is a broad spectrum antibiotic produced by *Streptoverticillium cinnamomeus* var. *terricola.* Aureofungin is effective only when it is mixed with copper sulphate (2 g per 1g of aureofungin sol). It controls Thanjavur wilt of coconut *(Ganoderma lucidurm),* citrus gummosis *(Phytophthora citrophthora),* chillies powdery mildew *(Oidiopsis* sp.) and grapes downy mildew *(Plasmopara viticola).* As seed treatment it controls *Helminthosporium oryzae* in rice seed.

(II) Streptomycin

It is produced by *Streptomyces griseus.* It controls bacterial diseases viz. citrus canker *(Xanthomonas citri),* blackarm of cotton *(Xanthomonas malvacearum),* and rice bacterial blight *(Xanthomonas campestris* pv. *oryzae).*

(III) Tetracycline

It is produced by *Streptomyces* sp and effective in controlling mycoplasma diseases like sandal spike, brinjal little leaf and gingelly phyllody. In combination with streptomycin (Streptocycline) it controls several bacterial diseases like rice bacterial blight and cotton black arm.

REFERENCES

1. Aharonson, N and U. Kafkafi, 1975. Adsorption of benzimidazole fungicides on montmorrilonite and kaolinite clay surfaces. J. Agric. Food Chem., **23**: 434-437
2. Al-Beldawi, A.S. and J.A. Pinokard, 1970. Control of *Rhizoctonia solani* on cotton seedlings by means of benomyl. Plant Dis. Reptr., **54**: 76-80
3. Allam, A.I., P.E. Schilling and J.B. Sinclair, 1969. Uptake of Vitavax (DCMO) by germinating cotton seeds. Phytopathology, **59**: 265-266
4. Bebtington, R.M., D.H. Brooks, M.J. Geoghegan and B.K. Snell, 1969. Ethirirmol: A new systemic fungicide for the control of cereal powdery mildews. Chem. Ind., **42**: 1513
5. Bhowmik, T.P. and N.V. Sundaram, 1971. Control of pearl millet smut with systemic fungicides. Plant Dis. Reptr., **55**: 87-88
6. Borum, D.E. and J.B. Sinclair, 1968. Evidence for systemic protection against *Rhizoctonia solani* with vitavax in cotton seedlings. Phytopathology, **58**: 976-980
7. Clark, E.M., P.A. Backman and R. Rodriques-Kabana, 1974. *Cercospora* and *Cercosporidium* tolerance to benomyl and related fungicides in Alabama peanut fields. Phytopathology, **64**: 1476-1477
8. Edgington, L.V. and G.L. Barron, 1967. Fungitoxic spectrum of oxathiin compounds. Phytopathology, **57**: 1256-1257
9. Edgington, L.V., B. Heinrich and F. Grossmenn, 1973. Bioassay and transcuticular movement of systemic fungicides. Pesticide Sci., **4**: 747-752
10. Edgington, L.V., K.L. Khew and G.L. Barren. 1971. Fungitoxic spectrum of benzimidazole compounds. Phytopathology, **61**: 42-44
11. Ellis, M.A. and J.B. Sinclair, 1975. Movement of methyl, 2-benzimidazole carbamate into soybean seeds after foliar application of benomyl. Phytopathology, **65**: 1387-1389

12. Erwin, D.C., 1973. Systemic fungicides: Disease control, translocation, and mode of action. Annu. Rev. Phytopathol, **11**: 389-422
13. Evans, E., 1971. Systemic fungicides in practice. Pesticide Sci., **2**: 192-196
14. Franke, W. 1967. Mechanism of foliar penetration of solutions. Annu. Rev. Plant Physiol., **18**: 281
15. Helling, C.S., D.G. Dennison and D.D. Kaufman,. 1974. Fungicide movement in soils. Phytopathology, **64**: 1091-1110
16. Horsfall, J.G., 1957. Principles of fungicidal action. Chronica Botanica, Waltham, Mass., U.S.A., p. 279
17. Horsfall, J.G. and A. E. Dimond, 1951. Plant Chemotherapy. Annu. Rev. Microbiol, **5**: 209-222
18. Kirby, A.H. M., 1972. Progress towards systemic fungicides. PANS, **18**: 1-33
19. Marsh, R.W., 1972. Systemic fungicides. Longman Press, London,. p. 321
20. Muchovej, J.J. and O.D. Dhingra, 1979. Benzene and ethanol for treatment of soybean seeds with systemic fungicide. Seed Sci. and Technol, **7**: 449
21. Muchovej, J.J. Q.D. Dhingra and P.R.S. Carvalho, 1980. Organic solvent as vehicle for systemic and non-systemic fungicides for vegetable seed treatment. Fitopathologia Fr., **5**: 337
22. Nene, Y.L. and P.N. Thapliyal, 1979. Fungicides in plant disease control. Oxford and IBH Publishing Co., New Delhi p. 507
23. Noveroske, R.L., 1975. Dowco 269: A new systemic fungicide for control of *Phytophthora parasitica* of tobacco. Phytopathology, **65**: 22-27
24. Raabe, R.D. and J.H. Hurlimann, 1974. Spray application of Dowco 269 to control *Pythium* root rot of carnation. Calif. Plant Pathol, **18**: 8-9
25. Sbragia, R.J., 1975. Chemical control of plant diseases - An exciting future. Annu. Rev. Phytopathol, **13**: 257-269
26. Torgeson, D.C., 1969. Fungicides — An advanced treatise Vol.I. Academic Press, New York. p. 561
27. Torgeson, D.C., 1969. Fungicides — An advanced treatise. Vol. II. Academic Press, New York
28. Vyas, S.C., 1984. Systemic fungicides. Tata McGraw-Hill Publishing Co. Ltd., New Delhi, p. 360.

20

DISEASE MANAGEMENT — BIOTECHNOLOGY (GENETIC ENGINEERING AND TISSUE CULTURE)

Plant Pathology has entered into an new era. Plant Pathologists have now learnt to use the modern tool of genetic engineering and tissue culture in management of crop diseases. It is now possible to isolate particular gene from one organism, insert them into the genome of another organism and make them to express at right time. This technology is known as 'Genetic Engineering'. Cells of plants can be cultured in special nutrient medium and whole plants can be regenerated from cultured cells. This technique of growing plants *in vitro* is called 'Tissue Culture'. The biological technology through which laboratory experiments are converted into practical utility materials, under field conditions is called 'Biotechnology'.

GENETIC ENGINEERING

(I) Vectors for Transfer of Genes

Genetic engineering has been utilised to manage virus diseases. For the transfer of genes to plants, vectors are needed in which the gene to be transferred will multiply several folds. The most effective gene vector so far developed for plant systems is the tumor-inducing (Ti) plasmid of *Agrobacterium tumefaciens* from which the tumor-inducing genes have been removed.

Cauliflower mosaic virus (CaMV) is the only plant virus with double-stranded DNA genome. As it has a DNA genome, it is used as a possible vector in introducing foreign genes into plant. It is possible to insert a non-viral gene into CaMV genome and obtain expression of the gene in the infected plant. The viral promotor regions from CaMV are effective for obtaining expression of other genes in plant cells. The genes to be expressed is now fused to a promotor element from CaMV and a gene of *A. tumifaciens.* They are then intorduced into the plants using *A. tumefaciens* Ti-DNA transformation.

cDNA construction: Messenger RNA is extracted and exposed to an enzyme reverse transcriptase which synthesises a complimentary single stranded DNA. The complimentary DNA (cDNA) is exposed to another enzyme, DNA polymerase, which produces the double stranded cDNA. The cDNAs are inserted into the plasmids of *A. tumefaciens.*

(II) Coat Protein Expression in Transgenic Plant

For successful infection of virus particles the capsid (coat) protein should be removed from viral RNA in many cases. If the host is made to synthesize capsid proteins in large amount, naked viral RNA formation will be negligible. The host capsid protein will encapsidate the RNA of the virus and prevent its multiplication. This will result in reduction and delay in symptom development.

Abel *et al.* (I) constructed a chimeric gene containing a cloned cDNA of the coat protein gene of tobacco mosaic virus (TMV), promoter from CaMV and a polyadenylation signal from the nopaline synthase gene of *A. tumefaciens.* The chimeric gene was introduced into tobacco cells on a Ti plasmid of *A. tumefaciens.* Then the cells were grown on nutrient media and the tobacco plants were regenerated from the transformed cells (tissue culture). The plants synthesized through genetic engineering expressed TMV mRNA and coat protein as a nuclear trait. The transgenic plants (the plants expressing transferred gene) expressing coat protein gene did not develop mosaic symptoms when inoculated with TMV.

By this method transgenic tobacco plants showing resistance to alfalfa mosaic virus and tobacco rattle virus have also been developed.

(iii) Satellite RNA Expression in Transgenic Plants

Satellite RNAs are associated with several viruses. They are packaged into virus particles along with the genomic RNAs of the helper virus. They are not part of the viral genome and have no obvious sequence relationships with the helper virus. The presence of the satellite RNA suppresses the disease severity in many hosts. Hence transgenic plants which express satellite RNA have been produced to manage virus diseases.

Gerlach *et.al.* (II) prepared synthetic plant gene constructions using cloned satellite RNA of tobacco ringspot virus cDNA segments. The inserts were fused to CaMV promoter gene and nopaline synthase gene of *A. tumefaciens.* They were then introduced into tobacco cells using *A. tumefaciens* Ti-DNA transformation. Transgenic plants expressed the synthesis of satellite tobacco ringspot virus and reduced the disease incidence. Similarly satellite RNA expressing tobacco plants against cucumber mosaic virus and tomato aspermy virus have been synthesized.

(iv) MIC RNA Expression in Transgenic Plants

A DNA copy is made of one or more sections of the viral genome that include the initiation codon for proteins vital to virus replication. The DNA copy is inserted into the host-cell genome. Cells then produce an 'antisera RNA' called mic RNA (mRNA-interfering complementary to 5' end of the gene). The mic RNA hybridizes *in vivo* with the viral mRNA blacking translation. The mic RNA is inserted into the plants using the Ti plasmid of *A. tumefaciens.* Plants regenerated from the transformed cells will be resistant to the particular virus. This possibility is also being exploited for the control of virus diseases (15).

TISSUE CULTURE

(i) Principles

In the tissue culture technique cells are cultured in nutrient media and whole plants are regenerated from cultured cells. Populations of cultured plant cells generally contain more genetic variability than whole plants. Plants regenerated from somatic cells are not genetically uniform but display significant genetic variability. This variability is called 'Somaclonal Variation'. Variation has been detected in the progeny of up to 15% regenerated plants. In general the somaclonal variations are due to single gene mutations and organelle gene mutations. Hence, it is possible to select new plants which retain all the favourable qualities of an existing variety while adding one additional trait, such as disease resistance. If the selected genes are recessive, no additional backcrossing is necessary. If the selected genes are dominant, only a few selfingsegregation-selection generations are needed to obtain homozygous plants. Hence screening cell cultures *in vitro* will be highly profitable to evolve disease-resistant varieties in a short time.

(ii) Practical Applications

Several disease-resistant plants have been evolved using somaclonal variation. Smith and Murakishi (1986) observed that 6 out of 370 tomato plants regenerated from calluses showed resistant to tobacco mosaic virus. Similarly late blight *(Phytophthora infestans)* -resistant potato plants and bacterial blight *(Xanthomonas campestris* pv. *oryzae)* resistant rice plants have been evolved.

The pathogen-produced toxins can be used to screen calluses (cultured cells) which may regenerate resistant plants. The toxins will kill the calluses, but the mutant toxin-resistant calluses will survive. The toxin-resistant calluses yield disease-resistant plants. Vidhyasekaran (19) obtained brown spot resistant rice plants using *Helminthosporium oryzae* toxin. Similarly *H. maydis* -resistant maize plants, *H. sacchari*-resistant sugarcane plants and *Phytophthora infestans*-resistant tobacco plants have been evolved.

Tissue culture has become an important tool to develop disease resistant varieties.

REFERENCES

1. Abel, P.P., R.S. Nelson, B. De, N. Hoffman, S.G. Rogers, R.T. Fraley and R.N. Beachy, 1986. Delay of disease development in transgenic plants that express the tobacco mosaic virus coat protein gene. Science, **232:** 738-743
2. Arciconi, S., M. Pezzotti and F. Damian,. 1987. In vitro selection of alfalfa plants resistant to *Fusarium oxysporium* f. sp. *medicaginis*. Theor. Appel Genet., **74**: 700-705
3. Beachy, R.N., D.M. Stark, C.M. Deom, M.J. Oliver, and R.T. Fraley, 1987. Expression of sequences of tobacco mosaic virus in transgenic plants and their role in disease resistance. Basic Life Sci., **41**: 169-180
4. Behnke, M. 1980. General resistance to late blight of *Solanum tuberosum* plants regenerated from callus resistant to culture filtrate of *Phytophthora infestans*. Theor. Appl. Genet., **56**: 151-152
5. Clark, G., D. Echholtz, R. Fraley and R.N. Beachy., 1987. TMV coat protein levels in transgenic tobacco plants affects the extent of cross protection. Plant Physiol, **83**: 128
6. Coleman, J., A. Hirashima, Y. Inokuchi, P. J. Green and M. Inouye. 1985. A novel immune system against bacteriophage infection using complementary RNA (mic RMA). Nature, **315**: 601-603
7. Dodda, J.H. and J.M. Jaynes. 1987. Crop plant genetic engineering: Science fixtion to Science fact. Outlook on Agriculture, **16**: 111-115
8. Evans, D.A. and W.R. Sharp, 1983. Single gene multations in tomato plants regenerated from tissue culture. Science, **221**: 949-951
9. Evans, D.A. and W.R. Sharp, 1986. Applications of somaclonal variation. Biotechnology, **4**: 528-532
10. Evans, D.A., W.R. Sharp and H.P. Medino-Filho. 1984. Somaclonal and gametoclonal variation. Amer. J. Bot., **71**: 759-794
11. Gerlach, W.L., D. Llewellyn and J. Haseloff, 1987. Construction of a plant disease-resistance gene from the satellite RNA of tobacco ringspot virus. Nature **328**: 802-805
12. Gustafson, J.P., 1984. Gene manipulations in plant improvement. Plenum Press, New York
13. Harrison, B. D., M.A. Mayo and D.C. Baulcombe, 1987. Virus resistance in transgenic plants that express cucumber mosaic virus satellite RNA. Nature, **328**: 799-802
14. Ling D.H., P. Vidhyasekaran, E.S. Borromeo, F.S. Zapata and T. W. Mew, 1985. *In vitro* screening of rice germplasm for resistance to brown spot disease using phytotoxin. Theor. Appl. Genet., **71**: 133-135

15. Matthews, R.E. F., 1987. The changing scene in plant virology. Annu. Rev. Phytopathol, **25**: 10-23
16. Nelson, R.S., P.P. Abel and R.N. Beachy, 1987. Lesions and virus accumulation in inoculated transgenic tobacco plants expressing the coat protein gene of tobacco mosaic virus. Virology, **158**: 126-132
17. Smith, S.S. and Murakishi, 1986. Nature of resistance of tomato somaclones to tomato mosaic vims and tobacco mosaic vims. Int. Congr. Plant Tissue Cult., 6 meet, p. 306.
18. Van Dun, C.M.P., J.F. Bol and L. Van Volten - Doting, 1987. Expression of alfalfa mosaic virus and tobacco rattle virus coat protein genes in transgenic plants Virology, **159**: 299-305
19. Vidhyasekaran, P., 1987. Use of biotechnology in evolving disease resistant rice varieties. Proc. Nat. Symp. Role of Biotechnology in Crop Protection, Kalyani, p. 6.

SUBJECT INDEX